CW01646132

Education and Equity

MyCopy

Emily S. Rudling • Sherridan Emery
Becky Shelley • Kitty te Riele
Jessica Woodroffe • Natalie Brown

Education and Equity in Times of Crisis

Learning, Engagement and Support

palgrave
macmillan

Emily S. Rudling
The Peter Underwood Centre
The University of Tasmania
Hobart, TAS, Australia

Sherridan Emery
The Peter Underwood Centre
The University of Tasmania
Hobart, TAS, Australia

Becky Shelley
The Peter Underwood Centre
The University of Tasmania
Hobart, TAS, Australia

Kitty te Riele
The Peter Underwood Centre
The University of Tasmania
Hobart, TAS, Australia

Jessica Woodroffe
The Peter Underwood Centre
The University of Tasmania
Hobart, TAS, Australia

Natalie Brown
The Peter Underwood Centre
The University of Tasmania
Hobart, TAS, Australia

https://doi.org/10.1007/978-3-031-18671-4

This Palgrave Macmillan imprint is published by the registered company Springer Nature Switzerland AG.
The registered company address is: Gewerbestrasse 11, 6330 Cham, Switzerland

Acknowledgements

We acknowledge and pay respect to traditional land ownership. This book was prepared in lutruwita, on the lands of the mouhenneener /muwinina people.

We would like to acknowledge and extend our thanks to those who made this book possible: school leaders and staff, education system officers, non-governmental organisation staff, and advocates. We are also grateful to Tess Crellin, Aishah Bakhtiar, and Damien Walker for creating the figures throughout the book and to Linda Bathish for her careful copy editing. Finally, our deepest thank you to colleagues at the Peter Underwood Centre for Educational Attainment for supporting this work. As always, we dedicate this to children and young people and their pursuit of learning.

Contents

About the Authors

Emily S. Rudling is a research fellow with the Peter Underwood Centre at the University of Tasmania. Her research explores the relationship between Australia and Asia and how this is reflected in Australian education policies, like intercultural competency.

Sherridan Emery is a research fellow with the Peter Underwood Centre at the University of Tasmania who is engaged in educational evaluation and research. She is an early career researcher exploring intersections between culture, well-being, education, and sustainability.

Becky Shelley is Deputy Director with the Peter Underwood Centre at the University of Tasmania. She has a discipline background in political science. With experience in complex social policy and projects for government, community, and peak bodies, her work intersects policy, research, and practice.

Kitty te Riele is interim Director of and Professor in the Peter Underwood Centre at the University of Tasmania. Her main research focus is on educational policies and practices that aim to improve opportunities to participate and succeed in school education for young people in disadvantaged communities.

Jessica Woodroffe is Coordinator of Partner Programs and a senior lecturer in the Peter Underwood Centre at the University of Tasmania. Her focus is on maintaining and developing formal educational programmes and partnerships that support further education pathways and which contribute to positive and sustained change in educational outcomes for young people—specifically for senior secondary students.

Natalie Brown is the Chair of Academic Senate at the University of Tasmania. She is engaged in the strategic direction of the university's engagement in academic qualification and advancement of education within the state of Tasmania. Further, she is Director of the National Schools Resourcing Board for the Federal Department of Education, Skills and Employment (Australia). For her work across these roles, she was awarded the 2020 Australian Council of Educational Leaders Preeminent Outstanding Education Leadership award.

List of Abbreviations

ABC	Australian Broadcasting Corporation
ABS	Australian Bureau of Statistics
ACARA	Australian Curriculum, Assessment and Reporting Authority
ACOSS	Australian Council of Social Services
AIHW	Australian Institute of Health and Welfare
CALD	Culturally and Linguistically Diverse
CoAG	Council of Australian Governments
COVID-19	Coronavirus Disease 2019
HBL	Home-Based Learning
ICT	Information and Communications Technology
ICPA	Isolated Children's Parents' Association
LMS	Learning Management System
NAPLAN	National Assessment Program—Literacy and Numeracy
NBN	National Broadband Network
NGO	Non-governmental Organisation
NSW	New South Wales (state of Australia)
OECD	Organisation for Economic Co-operation and Development
OER	Open Education Resource
PAYG	Pay As You Go
SELPS	Social and Emotional Learning Programs
SES	Socioeconomic Status
UK	United Kingdom
UN	United Nations
UNCRC	United Nations Convention on the Rights of the Child
UNFPA	United Nations Population Fund
UNESCO	United Nations Educational, Scientific and Cultural Organization
UNICEF	United Nations Children's Fund

USA	United States of America
USB	Universal Serial Bus
WHO	World Health Organization

List of Figures

CHAPTER 1

Facing a Crisis: Foregrounding the Future

Abstract Equity is a perennial concern in education. At a global level, this is recognised in the United Nations Sustainable Development Goals. In wealthy countries, educational inequality remains problematic, as evidenced in multiple sources such as the OECD Programme for International Student Assessment reports.

The COVID-19 pandemic radically altered children and young people's engagement with learning. The pandemic not only exacerbated existing inequities but created newly vulnerable groups: families who had lost jobs or who were struggling with combining their children's learning from home with the demands of working from home. The COVID-19 pandemic provided an impetus for innovation, for example in relation to the use of technology and enhanced interagency collaborations. The disruption caused by such crises is an opportunity to revisit questions of equity and education.

This chapter outlines our conceptual tools and provides an overview of the book.

Keywords COVID-19 pandemic • Equity • Education • Vulnerability • Learner rights • Social capital

E. S. Rudling et al., *Education and Equity in Times of Crisis*,
https://doi.org/10.1007/978-3-031-18671-4_1

Introduction

Equity is a perennial concern in education. International organisations, including the United Nations, have fluidly transmitted the possibility and promise of education for all (Robeyns, 2006). The right of the child to education has also been asserted with Article 29a of the United Nations Convention of the Rights of the Child (UNCRC), stating that "the education of the child shall be directed to the development of the child's personality, talents and mental and physical abilities to their fullest potential" (United Nations, 1989, p. 9).

In this book, we recognise the right of the learner to an equitable education. However, we examine equity in education through the lens of the COVID-19 pandemic to unpick what equity means in disaster and crisis. The COVID-19 pandemic is the largest recorded disruption to education globally (United Nations, 2020) with over a billion children affected (OECD, 2021). The closure of schools to manage the spread of COVID-19 pushed learning into the home environment and, in doing so, dismantled the supports schools offer learners, and blurred the once firm boundaries between school and home.

Despite some success in the second half of the twentieth century to improving access to education globally, access to education was in decline prior to the COVID-19 pandemic (Azevedo et al., 2020; World Bank, 2020). A "twin shock" to both education and economy (Azevedo et al., 2020, p. 2), children and young people are experiencing compromised access to education as well as broader social and economic stressors.

The pandemic has also raised a unique opportunity for a reset in education, one that, we hope, could go some way to redressing inequities. COVID-19 school closures have spurred innovations in the digital provision of learning as well as creating new challenges to equitable access to education while learning from home (Sahlberg, 2021; Zhao & Watterson, 2021). The role of schools in remedying educational inequities has also altered: now, more than ever the home environment and socio-economic context mediate learner access to education.

Our intention is not to attempt a comprehensive review of global experiences of the pandemic. Instead, we analyse the effects of the COVID-19 pandemic on the provision of education to children and young people to understand some of the critical challenges that crises present for equity in education systems. It is in this context that we explore what we can learn from the global pandemic to reset and reconsider education and the role of schools, and outline ongoing implications for schools and society, not only for times of crisis but also for enhancing educational equity more broadly.

There is urgency to ensure education is equitable for learners during and beyond crises like the COVID-19 pandemic (Sahlberg, 2021). Despite the reality that disaster and crisis are recurrent features of the global landscape, there is little understanding of how such events impact on learners. Socio-economic disadvantage and privilege affect the extent to which a learner is impacted by crisis; and crisis and disaster have the potential to create cohorts of newly vulnerable learners. For example, learning loss as a result of disaster and crisis have long-term impacts on individual and national wellbeing and productivity (Azevedo et al., 2020). As we will discuss in Chap. 3, historical crises and disasters teach us that periods of heightened stress have the potential to affect the entire life course of a child (Yoshikawa et al., 2020). The pursuit of educational equity is thus enmeshed within broader socioeconomic conditions.

Throughout this book, we draw on international literature to underline our claim that educational equity is of global concern. We focus on literature from nations with similar education systems: Australia, the United Kingdom, New Zealand, Japan, and the United States of America. We recognise that circumstances in developing countries create different challenges, which are largely beyond the scope of this book.

The pandemic is not the only crisis to have affected the world in the early 2020s. Earthquakes, floods, and bushfires as well as ongoing and new military conflict and wars have affected children's lives—including learning, development, and access to schooling—in many countries. These events are beyond the remit of this book but are a reminder that the relevance of our contention—that equity in education matters—extends beyond the COVID-19 pandemic to other crises, such as those triggered by climate change.

Conceptual Tools

Educational outcomes and experiences (as well as other measures such as life expectancy, physical and mental health) correlate strongly with geography, gender, race, and poverty (Hill Collins & Bilge, 2020). Much research shows how educational inequality relates to socioeconomic status. This body of work, which we draw on across the book, provides important insights into the significance of equity to educational outcomes. However, disadvantage is multidimensional. It is not just about a lack of material resources, but also lack of access to rights and services, and "the inability to participate in the normal relationships and activities, available to the majority of people in a society" (Levitas et al., 2007, p. 25). Disadvantage is also complex, with people exposed to multiple dimensions of disadvantage at risk of facing deep exclusion (Levitas et al., 2007) highlighting the layering effects of patterns of power, discrimination and inequality that create compounding barriers in a variety of ways (Hill Collins & Bilge, 2020). To acknowledge this complexity, and the intersectional nature of educational equity, we draw on three conceptual tools to build a theoretical framework.

First, tools provided by Urie Bronfenbrenner are useful. A central theme across his seven decades of work on human development is that "children must be nurtured and educated to be able to maintain and strengthen their society" (Hamilton & Ceci, 2005, p. 284). It was Bronfenbrenner's view that societies should be judged according to "the concern of one generation for the next" (Hamilton & Ceci, 2005, p. 284). Bronfenbrenner's (1977) ecological system model places the child at the centre and highlights the multiple layers of systems that sit around the child—each offering loci for influences that can support or inhibit children's learning, wellbeing, and development. This is of particular importance in the context of crises, such as the COVID-19 pandemic, when families, schools, communities, employers, governments, and non-government organisations all swing into action to respond. While many of these responses are not explicitly targeted towards children and young people, they inevitably impact them.

Secondly, theories of capital are valuable. Social capital, according to Putnam (2000) and Bourdieu (1984), refers to the advantages generated by connections between people and amongst social groups. Through its potential for helping people gain access to resources, Putnam (2000) argues that social capital also acts as a public good, and benefits can reach the larger community. School can be a key site for the building of children's social capital according to Guilfoyle et al. (2011), as a place where the formation of networks of skills, knowledge and relationships occurs. Public health research identifies positive associations between social capital and health (Helliwell & Putnam, 2004; Uphoff et al., 2013).

Social capital is also a cornerstone of Pierre Bourdieu's theories of capital (1984) which further encompass economic and cultural capital. Economic capital refers to money, property and other financial assets people use to acquire resources. Cultural capital is conceptualised as taking three forms. First, institutionalised cultural capital refers to educational attainment and qualifications, and relevant here is the cultural capital located in schools and educational institutions. Second, objectified cultural capital relates to the possession of cultural goods and artefacts. Third, Bourdieu's concept of embodied cultural capital, refers to people's values, skills, knowledge, and tastes. Social and cultural capital have been widely employed in relation to education (Bourdieu & Passeron, 1990) and are helpful in explaining the social reproduction of advantage and disadvantage. These theories provide us with tools to consider the differential access to resources that supported students during the pandemic, through the shift to learning at home rather than at school.

Lastly, crises tend to involve a re-ordering of people's priorities, especially when day-to-day survival is suddenly under threat. In recognition of this, we draw on Abraham Maslow's (1954) hierarchy of needs which places the child's security and physiological needs as the basis for healthy development, together with broader psychological needs of love, belonging and esteem. These, according to Maslow, precede the attainment of self-fulfilment needs such as self-actualisation. Noltemeyer et al. (2021) further unpack the antecedent/subsequent elements of Maslow's theory, distinguishing "deficiency needs", namely children's food, shelter, health, and psychological needs, from "growth needs", encompassing academic work, achievement, and self-esteem. In arguing for the necessity of children's basic (or "deficiency") needs being met as precursors to their learning (or "growth") needs, Noltemeyer et al. (2021) help to explain why children in families with higher incomes and capital resources are typically well placed to have the motivation to pursue higher order growth needs.

Maslow's work has met with important critique. By framing his theory as focused on human motivation, there is a risk that Maslow places too much emphasis on the responsibility of individuals for the decisions they make (Noltemeyer et al., 2021). This in turn, can be interpreted as blaming individuals for their circumstances without recognising broader factors like precarity, gender, or race (Hill Collins & Bilge, 2020). We nevertheless argue that the idea of a hierarchy of needs is both commensurate with both ecological systems theory and theories of capital and an important addition for the analysis in this book. When informed by sociological understandings of how social reproduction of advantage and disadvantage conditions a landscape (or ecological system) of inequality in education, Maslow's hierarchy offers insight into the constellation of environmental factors that are central to children's development, and which are particularly thrown into disarray in a crisis.

Overview of the Book

This chapter sets the scene for the book, outlining how we examine the connection between equity and education in the context of crisis, with a particular focus on the COVID-19 pandemic. The book is based on an extensive analysis of international scholarly and grey literature, alongside primary data from key informants in Australia and New Zealand, representing schools and education systems as well as relevant non-government organisations and allied professions. An array of rich formal and informal networks and connections are drawn upon to develop the capabilities of children and young people—these too have been impacted by COVID-19 related school closures and general geographical "lockdowns". The longer-term effects of these choices upon learning and development are unfurling, and emerging data from the current crisis as well as experiences from past crises provide insight into how these are already affecting equity in education. Our argument here is that crises, such as the COVID-19 pandemic, can exacerbate (and alter) inequalities, as well as present opportunities for action through sharpened attention for the need to address educational inequities. To address this, we need analyses from around and across the world, such as presented in this book.

In Australia, in the early days of the COVID-19 outbreak, the Federal Minister for Education sought expert advice from within the education field, leading to five rapid response reports [1]. We produced one of these reports (see: Brown et al., 2020). What we learnt from that early work, and from ongoing reflection as education and health responses to the pandemic unfolded, suggested the need for a longer piece, enabling deeper analysis of the educational equity implications of the pandemic. We consider research against different points in time—from interviews conducted early in the pandemic to later in the pandemic, and beyond. For example, at early points in the COVID-19 pandemic, feelings of stress in the education system were evident in conversations with our research participants and correspondents. Sympathetic to that pressure, we continued our inquiry six months later and found that while the initial sense of being overwhelmed had abated, new and underlying educational issues were emerging, which are discussed throughout this book.

[1] Available here: https://www.dese.gov.au/about-us/announcements/new-research-examine-potential-impact-remote-learning.

Employing a pragmatic approach (Tashakkori & Teddlie, 2008), the key methods and sources of information for our primary data are interviews and email correspondence with key informants. Purposive sampling methods were used to access key informants from national networks. The data is cross-sectoral, which in this study means that it includes several school sectors operating in Australia (e.g., government, Independent, Catholic, and flexible education systems as well as early childhood education and care). It also goes beyond the education system to other key stakeholders who, as per Bronfenbrenner's model (see above) influence learning and development.

Table 1.1 introduces the primary data collected during the two rounds of interviews. Five participants provided insights in both rounds.

The categories in the table above are used throughout the book to indicate the source of quotes. In addition, we contributed to two other research projects on the impacts of the COVID-19 pandemic on learning and on children's experiences:

- Children's experiences and conceptions of learning at home through COVID-19 (see Bourke et al., 2021).
- Learning through COVID-19 (see McDaid et al., 2020, 2021a, 2021b; Tomaszewski et al., 2022; Plage et al., 2022)

Throughout the book, we use publications from those projects as well as our own report (Brown et al., 2020) as a source of data, except where the data has not previously been published. As noted earlier, findings presented in this book are based on analysis of primary data sets as well as international scholarly literature and a wide range of grey literature,

Table 1.1 Overview of Research Participant Sample

Category	*Number*
School-based staff (i.e., teachers and support staff)	9
School-based leaders (i.e., principals and assistant principals)	7
Early childhood education and care (cross-sectoral)	3
Education system (cross-sectoral)	6
Advocacy Groups and Services (e.g. from NGOs, foundations, child advocacy bodies, and peak bodies)	10
Total number	**35**

including reports, media, and published stories by children and young people.

As parents/carers and grandparents, we wrote the book at a time when those closest to us were experiencing the impacts of lockdowns and disconnection from physical schooling. We acknowledge and are grateful to those who participated in our data collection at a time when they, like the rest of the world, were also facing challenges and pressures.

As this chapter has already indicated, broader structural inequalities play a key role in education, as in other of our society. These are addressed in detail in Chap. 2. We then take a birds' eye view in Chap. 3 of lessons from crises for educational equity: first based on literature about past crises, and then in relation to the COVID-19 pandemic.

The high-level analyses from Chaps. 2 and 3 are taken up in more detail in the three subsequent chapters. Chapter 4 examines inequality in relation to school-based learning, in particular analysing how this evolved during COVID-19 induced "lockdowns" when most students were "learning from home". Chapter 5 widens the scope to the broader role of schools, particularly in relation to wellbeing and material supports for vulnerable students. Chapter 6 turns to changing pedagogies, and in particular the use of a range of technologies for learning, as highlighted through the pandemic.

The final chapter brings together the findings and arguments from Chaps. 2–6 in relation to the conceptual tools established in this chapter. Here, we introduce our model of educational equity based on an adaption of Bronfenbrenner's ecological model. It provides implications for policy and practice, to improve educational equity not only in relation to the COVID-19 pandemic, but for the longer term.

Conclusion

This book is concerned with re-imagining the nature, form, and content of educational equity in the face of complex enmeshed local, national, regional, and global crises like the COVID-19 pandemic. The intersection of local and global problems demands a rethinking of the actual underlying principles and practices of education. The explosion of interest in online learning, wellbeing, and other concerns has left unquestioned whether the social safety nets conceived during the crisis should be adopted long term. The phenomenon of educational inequity is not new, but the COVID-19 pandemic has precipitated an educational crisis with devastating consequences. Yet, there is potential for a transformation of the politics of education in the wake of the pandemic and the realisation of the interconnectedness of inequity and the intensity of its impacts. Conceptual and institutional innovations regarding education and the nature of school itself are emerging. To the extent that the education system has been part of the maintenance and reproduction of systemic inequalities of outcome, there is now an opportunity when faced with a distinctive and novel crisis to repattern its practices. The book lays the basis for the advocacy of a model of equity in education.

As will be addressed in Chap. 7, our book builds to a model for educational equity. We put the learner at the centre of the educational enterprise and argue that access to learning is filtered by family context, school, and school system. These factors influence, and are influenced by, material basics and emotional wellbeing, and ultimately, affect access to learning.

It is imperative that there is a concerted effort to create circumstances for educational policy that is framed by an assumption of respect for children's individual capacities, their ability to learn and their potentialities in times of crisis and disasters, as well as in times of normality. Understandings about supporting learning for vulnerable children and young people have been sharpened due to the pandemic. We outline ongoing implications for schools and learning, not only for times of crisis but also for enhancing educational equity more broadly to reconsider education and the role of schools.

COVID-19 is not the first crisis affecting children's learning and wellbeing, and it will not be the last. There may be other pandemics, and there will be environmental crises wrought by climate change: earthquakes, floods, and bushfires.

References

Azevedo, J. P., Hasan, A., Goldemberg, D., Iqbal, S. A., & Geven, K. (2020, June 18). Simulating the potential impacts of COVID-19 school closures on schooling and learning outcomes: A set of global estimates. *World Bank Group: Education global practice*. Retrieved February 2, 2021, from https://www.worldbank.org/en/topic/education/publication/simulating-potential-impacts-of-covid-19-school-closures-learning-outcomes-a-set-of-global-estimates

Bourdieu, P. (1984). *Distinction: A social critique of the judgement of taste*. Harvard University Press.

Bourdieu, P., & Passeron, J. C. (1990). *Reproduction in education, society and culture*. Sage University Press.

Bourke, R., O'Neill, J., McDowall, S., Dacre, M., Mincher, N., Narayanan, V., Overbye, S., & Tuifagalele, R. (2021). *Children's informal learning at home during COVID-19 lockdown*. New Zealand Council for Educational Research and Massey University. Retrieved October 1, 2021, from https://www.nzcer.org.nz/system/files/NZC_Children%27s%20informal%20learning%20Final%2023-8-21.pdf

Bronfenbrenner, U. (1977). Toward an experimental ecology of human development. *The American Psychologist, 32*(7), 513–531. https://doi.org/10.1037/0003-066X.32.7.513

Brown, N., te Riele, K., Becky, S., & Woodroffe, J. (2020). *Learning at home during COVID-19: Effects on vulnerable young Australians* (Independent Rapid Response Report). Peter Underwood Centre for Educational Attainment, University of Tasmania. Retrieved May 26, 2020, from https://www.utas.edu.au/__data/assets/pdf_file/0008/1324268/Learning-at-home-during-COVID-19-updated.pdf

Guilfoyle, A. M., Banham, V., Cavazzi, T., & Napolitano-Lincoln, R. (2011). An ecological approach to supporting children after parental separation: Resilience, social capital and sense of community in the school classroom. *International Journal of Learning, 18*(1).

Hamilton, S., & Ceci, S. (2005). Afterword. In U. Bronfenbrenner (Ed.), *Making human beings human: Bioecological perspectives on human development* (pp. 283–284). SAGE Publishing.

Helliwell, J. F., & Putnam, R. (2004). The social context of well-being. *Philosophical Transactions of the Royal Society London, 359*, 1435–1446.

Hill Collins, P., & Bilge, S. (2020). *Intersectionality* (2nd ed.). Wiley.

Levitas, R., Pantazis, C., Fahmy, E., Gordon, D., Lloyd-Reichling, E., & Patsios, D. (2007). *The multi-dimensional analysis of social exclusion*. University of Bristol.

Maslow, A. (1954). *Motivation and personality*. Harper & Row.

McDaid, L., Cleary, A., Robinson, M., Johnstone, M., Curry, M., Ridgway, T., & Western, M. (2021a). *What can be done to maximise educational outcomes for children and young people experiencing disadvantage?* (Pillar Report No. 3). Institute for Social Science Research, University of Queensland. Retrieved June 16, 2021, from https://issr.uq.edu.au/LTC-Pillar3

McDaid, L., Povey, J., Tomaszewski, W., Cleary, A., Plage, S., Cook, S., Xiang, N., Zajac, T., Ridgway, T., & Western, M. (2021b). *What do children, young people and families say about needs, and impact of COVID-19?* (Pillar Report No. 2). Institute for Social Science Research, University of Queensland. https://issr.uq.edu.au/files/21372/Ramsay%20Pillar%202_Final.pdf

McDaid, L., Tomaszewski, W., Staton, S., Povey, J., Thorpe, K., Vaithianathan, R., Ridgway, T., Diffey, B., & Western, M. (2020). *Who are the students most at risk of falling being in their learning?* (Pillar 1 Report). Institute for Social Science

Research, University of Queensland. Retrieved June 16, 2021, from https://issr.uq.edu.au/files/17185/Ramsay%20Pillar%201_Full%20Report.pdf

Noltemeyer, A., James, A. G., Bush, K., Bergen, D., Barrios, V., & Patton, J. (2021). The relationship between deficiency needs and growth needs: The continuing investigation of Maslow's theory. *Child & Youth Services, 42*(1), 24–42. https://doi.org/10.1080/0145935X.2020.1818558

Organisation for Economic Co-Operation and Development. (2021). *The state of education one year into the COVID-19 pandemic* [Webinar]. Retrieved April 4, 2021, from https://oecdedutoday.com/oecd-education-webinars/

Plage, S., Cook, S., Povey, J., Rudling, E., te Riele, K., McDaid, L., Western, M., Connection, connectivity and choice: Learning during COVID-19 restrictions across mainstream schools and Flexible Learning Programmes in Australia. *Australian Journal of Social Issues* 0, 0. 1-20. https://doi.org/10.1002/ajs4.228

Putnam, R. (2000). Social capital: Measurement and consequences. *Canadian Journal of Policy Research, 2*(1), 41–51.

Robeyns, I. (2006). Three models of education: Rights, capabilities and human capital. *Theory and Research in Education, 4*(1), 69–84. https://doi.org/10.1177/1477878506060683

Sahlberg, P. (2021). Does the pandemic help us make education more equitable? *Education Research Policy and Practice, 20*, 11–18. https://doi.org/10.1007/s10671-020-09284-4

Tashakkori, A., & Teddlie, C. (2008). Quality of inferences in mixed methods research: Calling for an integrative framework. *Advances in Mixed Methods Research, 53*(7), 101–119. https://doi.org/10.4135/9780857024329.d10

Tomaszewski, W., Zajac, T., Rudling, E., te Riele, K., McDaid, L., & Western, M. (2022). Uneven impacts of COVID-19 on the attendance rates of secondary school students from different socioeconomic backgrounds in Australia: A quasi-experimental analysis of administrative data. *Australian Journal of Social Issues, 00*, 1–20.

United Nations. (1989). *Convention on the Rights of the Child*. Office of the United Nations High Commissioner for Human Rights. Retrieved April 19, 2021, from https://www.ohchr.org/Documents/ProfessionalInterest/crc.pdf

United Nations. (2020). *Policy brief: Education during COVID-19 and beyond*. Retrieved November 16, 2020, from https://www.un.org/development/desa/dspd/wp-content/uploads/sites/22/2020/08/sg_policy_brief_covid-19_and_education_august_2020.pdf

Uphoff, E. P., Pickett, K. E., Cabieses, B., Small, N., & Wright, J. (2013). A systematic review of the relationships between social capital and socioeconomic inequalities in health: A contribution to understanding the psychosocial pathway of health inequalities. *International Journal for Equity in Health, 12*(1), 54. https://doi.org/10.1186/1475-9276-12-54

World Bank. (2020). The COVID-19 pandemic: Shocks to education and policy responses. https://inee.org/sites/default/files/resources/148198.pdf

Yoshikawa, H., Wuermli, A., Britto, P. R., Dreyer, B., Leckman, J. F., Lye, S. J., Ponguta, L. A., Richter, L. M., & Stein, A. (2020). Effects of the global coronavirus disease-2019 pandemic on early childhood development: Short- and long-term risks and mitigating program and policy actions. *The Journal of Pediatrics, 223*, 188–193. https://doi.org/10.1016/j.jpeds.2020.05.020

Zhao, Y., & Watterson, J. (2021). The changes we need: Education post COVID-19. *Journal of Educational Change, 22*, 3–12. https://doi.org/10.1007/s10833-021-09417-3

CHAPTER 2

Educational Vulnerability During COVID-19

Abstract The COVID-19 pandemic was initially regarded as something that didn't discriminate and to which we, collectively, were all vulnerable. Yet, its impacts—health, economic, and educational—have been experienced disproportionately by groups that experience poverty and marginalisation. This chapter highlights matters of inequality and explores the ways in which they manifest in educational inequity and are exacerbated during disaster and crisis. People who experience racism, discrimination (on the basis of gender, culture, disability status), as well as poverty and precarity were more likely to suffer direct and indirect consequences of the pandemic. The concept of vulnerability that this chapter engages with highlights structural issues in society which impact access to education during the pandemic and more broadly during times of crisis. The analysis reveals the need for specific programs, policies, and institutions to address structural and systemic injustice that influence educational vulnerability, in particular by addressing racial and gendered injustice through a focus on economic justice and childhood poverty.

Keywords COVID-19 pandemic • Vulnerability • Inequity • Education • Poverty • Race • Gender • Privilege • Social systems • Precarity

E. S. Rudling et al., *Education and Equity in Times of Crisis*,
https://doi.org/10.1007/978-3-031-18671-4_2

Introduction

During crises, we need to be concerned about the direct and indirect consequences of these events for children and young people and their educational attainment. Historical legacies of gender, race, and global divisions of labour set the context in which the COVID-19 pandemic has evolved. During crises, particularly those that result in the closure of key services for families and children in need, the effects of structural and systemic inequalities are amplified. Educational systems, globally, serve as an avenue to support broader social policy goals. Bronfenbrenner's (1977, 1995) ecological systems theory describes the dependency of living creatures on their surroundings—the ecological systems—and stresses on how human development is influenced by the different types of environmental systems. Ecological systems theory explains the influence of the interaction between children and their environments on their growth and development (Bandura & Walters, 1977; Bronfenbrenner, 1977). The more encouraging and nurturing these relationships and places are, the better the child outcomes. This chapter aims to create a sense that what people do matters, rather than providing a narrative about the problem of inequality. During the COVID-19 pandemic, families and care givers, services and teachers pivoted to try to sustain children's engagement in learning despite the challenges the pandemic presented.

For many throughout COVID-19, the domestic/family unit became the environment in which education was delivered and hence we consider how broader family and societal contexts affected the delivery of education during this time. Socioeconomic realities influence the context within which families strive to support their children's learning and broader wellbeing outcomes. Disparities in educational outcomes are a loss of human potential (which is concerning in and of itself) as well as a contravention of children's rights under the United Nations *Convention on the Rights of the Child* (1989) (Azevedo et al., 2020). A complex interplay of risk factors, service disruptions and the diversion of resources amplifies educational vulnerability.

The concept of vulnerability is not abstract, neutral, or objective. It is relational and social, and it is therefore fruitful to examine vulnerability in relation to specific contexts (Butler, 2016). There are justified objections to the term vulnerability. Such identification may lead to increased stigmatisation. Another objection is that the term inadequately conceives people's agentic power. For our purposes we recognise that vulnerability is not a fixed field. It is not a "subjective disposition, but a relation to a field of objects, forces, and passions that impinge upon or affect us in some way" (Butler, 2016, p. 16). Butler calls attention to a dual dimension in which vulnerability involves acting and being acted upon (2016, p. 6). As the book reveals, faced with difficult choices and under unfair constraints, there are many instances of children and young people who are vulnerable engaging in learning, both formal and informal, during the pandemic. Clearly, vulnerability does not determine outcomes, however, it does create additional issues that need to be recognised if equity it to be pursued.

The global pandemic has disrupted the lives of children and young people: exposing and worsening the inequalities and injustices that exist between countries just as it has within countries. These COVID-19 pandemic related inequalities are shaped by processes of marginalisation which, before and during the pandemic, influence well-documented educational, social and health inequalities (Jani, 2020; Marmot & Wilkinson, 2005; Sharma et al., 2020). The policy responses in Australia and elsewhere during the COVID-19 pandemic is changing school, family and community life in ways which have the potential to damage children's wellbeing and development intergenerationally. For example, increases in poverty and food insecurity, limited access to community-based health and family support services, and heightened stress can affect not only the entire life course of the child, but also future generations through physiologic, psychological, and epigenetic changes occurring in-utero and during early development that may decelerate the gains made since the turn of the century (Yoshikawa et al., 2020, p. 191).

The concept of vulnerability is useful to understand and reveal the way a global crisis exacerbates inequities in education, creates new areas of need, and generates an opening for a renewed appreciation of our shared humanity. Hopefully, elucidating the ways in which the crisis manifests in the lives of children and young people renews a commitment to the collective choices that will improve educational outcomes over the life course. This chapter examines broad social, cultural, and economic issues that influence the lived experiences of children and families at home and school to examine what claims we might collectively make of the structures and agencies that serve children in order to address educational vulnerability now and into the future. For clarity, various structural inequalities are discussed separately below. However, we recognise (as outlined in Chap. 1) that in people's lived experience these intersect in ways that can create additional complexities and deep exclusion (Hill Collins & Bilge, 2020; Levitas et al., 2007).

Gender and Identity

Gender related educational vulnerability presents differently depending on the socioeconomic context of a nation, as well as the religious and cultural beliefs of that nation. The COVID-19 pandemic may lead to an increase in educational inequity, with girls expected to perform additional household chores and, in some countries, agricultural work (International Labour Organisation and United Nations Children's Fund, 2020).

Girls may also be significantly disadvantaged in education due to sexual and reproductive health issues and because some girls have less time available for study and extra-curricular learning because of their disproportionate burden of household work. In Australia children's time use is known to be associated with gender. Boys tend to average more time in leisure activities and less time doing housework than girls and these gender differences become more pronounced as children grow older (Australian Institute of Family Studies [AIFS], 2014, p. 52).

Policy responses to the shift to learning at home assumed/required parents and caregivers to have resources, time, and capabilities to support learning at home. Craig and Churchill (2021, pp. 68–69) conducted a survey of adult Australian gender differences in objective time spent on activities, and in subjective feelings about this, during the pandemic. Mothers' unpaid work time went up, with a large proportion of them reporting feeling that they were doing much more than their fair share and were even more dissatisfied with how they and their partners shared unpaid work and care during COVID-19 (Craig & Churchill, 2021). Time use and gendered roles were reflected in interview data, including the following comment from a female secondary school student:

> You can get distraction. […] I pay more attention when I'm in class than online classes. … Things like my brothers, they get really annoying as well. […] When they are finished their classes, they just run around the house and I can't really focus. […] And since I was home, when my mum is cooking, I feel bad sitting in my room and doing my work, so I sometimes go and help her with some stuff. (as cited in McDaid et al., 2021, p. 30)

Women's capacity to nurture learning in the home and to contribute to household income has been shown to be particularly vulnerable during the pandemic. Globally gender hierarchies manifest in the form of measurable inequities in employment, earnings, and time use. Unlike during earlier economic recessions, pandemic employment losses have been larger for women than for men (Foley & Williamson, 2021). Two drivers have been found that would explain why: (1) lockdown measures and fear of infection have mostly affected sectors/occupations with a high concentration of female workers; (2) the closures of schools and early childhood education and care centres, and the implementation of remote learning have increased home-based childcare needs forcing many parents, particularly mothers, to choose between keeping their jobs or taking care of children (Fabrizio et al., 2021). These problems are particularly acute for single parents who work in essential jobs in areas where schools and early childhood centres closed and instead shifted to learning at home. Many of these challenges have more severe consequences for women of colour and lower-income women, especially those who have needed to continue working outside of the home in jobs considered essential (Cohen & van der Meulen Rodgers, 2021, p. 2).

Disturbingly, family violence (including inter-partner violence, child abuse and elder abuse) escalate during and after large scale crises (Bradbury-Jones & Isham, 2020, p. 2047). For example, the Ebola crisis had a significant impact on girls, with increases in teen pregnancy, sexual violence, and disproportionate school dropout and marginalisation (Yoshikawa et al., 2020, p. 189). During the COVID-19 pandemic, women and girls also have experienced an escalation of violence. The United Nations has described violence against women and girls during COVID-19 as the "shadow pandemic" (United Nations Entity for Gender Equality and the Empowerment of Women, 2020). Restrictions on movement may shut off avenues of escape and help-seeking, as well as impact ways of coping.

International research also suggests heteronormativity and cisnormativity in practice and policy-level responses to large-scale disasters systematically ignores the needs of LGBTQI+ populations (Salerno et al., 2020). LGBTQI+ youth are likely to be affected because physical isolation may simultaneously result in confinement to unsupportive home environments and isolation from developmentally important social connections, and identity-based resources; circumstances which are associated with compromised mental health and likelihood to engage in health risk behaviours, such as substance use or self-harm (Fish et al., 2020).

The issues briefly outlined above are vitally important from an equity and justice perspective. They are also critical to child development and wellbeing outcomes.

Racism and Discrimination

Globally, the COVID-19 outbreak has uncovered an uncomfortable propensity towards racism, prejudice, and intolerance (Devakumar et al., 2020). The World Health Organisation changed identification of strains of the virus from country-of-origin terminology to the Greek alphabet in response to these disturbing issues. Although the form of discrimination might vary, they contribute to inequalities across the spectrum of educational, economic, and wellbeing outcomes. In some societies, they are based on race or ethnicity; in others, colour, caste, religious beliefs, Indigeneity, or someone's migratory status (Chakraborty, 2020). The reality of a two–four times increased mortality risk from COVID-19 for minority groups has brought to light social and structural injustices that have existed for centuries and are derived from the same intersecting systems of oppression (Devakumar et al., 2020). The race-based social, economic, and health inequities that have been evident during the COVID-19 pandemic have not emerged randomly nor passively; the field of epidemiology is increasingly recognizing that the system of racism is more than a passive backdrop; racism is a dynamic functioning epidemic that has converged with the COVID-19 pandemic to accelerate exposure, disease, and mortality (Poteat et al., 2020). In education, it has underscored the necessity to ensure inclusion of perspectives, experiences, and voices of people of colour and specific groups when crafting and evaluating educational policies and practices (Dixson & Rousseau Anderson, 2018).

In Australia, legacies of invasion, colonisation, dispossession, and racism have been absorbed across and within generations of Aboriginal and Torres Strait Islander people with impacts clear within 'Overcoming Indigenous Disadvantage Indicators' (Productivity Commission, 2020). Concerningly, the indicator for year 1–10 school attendance regressed in 2020 (Productivity Commission, 2020). Griffiths et al. (2016, p. 16) note that there is an acceptance of "truth" about the negative effects of the period of colonisation upon Indigenous Australians but a persistence of colonial views that continues to play out overtly and systematically in contemporary policies (Cross-Townsend, 2011). Burgess et al. (2022) advocate for culturally responsive schooling and a culturally sustaining approach to everyday teaching practice to support decolonising Indigenous education.

Rates of Aboriginal and Torres Strait Islander children in out-of-home care have almost tripled in the past 15 years (60 per 1000 children in 2018–19) (Productivity Commission, 2020, p. xxiii). In June 2020 56.3

per 1,000 Indigenous children were in care, compared to 5.1 for non-Indigenous children (Australian Institute of Health and Welfare [AIHW], 2021c). Whilst the youth detention rate has decreased over the past decade, it is still 22 times the rate for non-Indigenous youth. During the COVID-19 pandemic, in the June quarter of 2020, 23 per 10,000 young Aboriginal or Torres Strait Islanders aged 10–17 was in detention, compared with 1.3 per 10,000 young non-Indigenous Australians (AIHW, 2021a, p. 12). Just under half (48%) of all young people in detention on an average night in the June quarter 2020 were Aboriginal or Torres Strait Islander young people. Yet Aboriginal and Torres Strait Islanders Australians make up just six percent of the Australian population aged 10–17 (AIHW, 2021a, p. v). These statistics are not only an indictment on Australian society in general, but during the pandemic also meant further levels of risk and exclusion for these Aboriginal and Torres Strait Islander children and young people. For example, staff in our research gave examples of foster families no longer able to look after their foster children, creating even more instability and trauma for these children in care:

> What I am aware of, I don't know if this is official, a number of our foster families are saying, because they've got their own kids at home now, they don't want foster kids. And they're saying, 'you have to take them, we don't want them, we've got our own three kids at home all the time now and that's enough, we can't manage two more'. So there are some young people who are being displaced. (Education system staff as cited in Brown et al., 2020, p. 15)

Notwithstanding the above, there are many instances where local responses have worked to combat historical legacies. One such example is described by Walker et al. (2021) who reveal the strength of local Indigenous communities and the importance of connection to culture and country in "place-based" responses to managing COVID-19 highlighting the power of both Indigenous self-determination and resilience (Walker et al., 2021).

Digital exclusion formed an additional barrier to education during the pandemic. Only 63% of Indigenous people in Australia have access to internet at home (Australian Bureau of Statistics [ABS], 2012). When online learning became the norm, a lack of internet access and digital technology was a key barrier to participation. Social distancing measures and social isolation have severely limited opportunities for social engagement among young people. For many Indigenous young people, these connections are being maintained and sustained using modern technologies to access mental health services and technologies (including apps), call lines and information as well as to support access to friends and families (Walker et al., 2021).

Discrimination intersects with other dimensions of identity and experience—culture, ethnicity, class, disability status. Culturally and linguistically diverse (CALD) and migrant or refugee families and children have experienced difficulties with online learning in Australia. One Advocacy Groups and Services staff member from a non-government organisation (NGO) explained:

> In a lot of multicultural families, many families didn't really even understand that school was stopping and didn't even know that they needed to collect educational packages.... A lot of the information, the COVID information, was not translated into different languages, or within the education department. One of our workers actually advocated for the need for the COVID information on their website to be translated. And now it is, which is great. Some parents who are not really engaged with education anyway, they're not very likely to be going to school to collect learning packs. So, if the parents aren't engaged, then it's unlikely that the child/some children even actually received any learning packs.

One student (female, secondary school) evaluated the relative capacities for her family members to support her learning at home: "My mum doesn't understand English, my dad's [and] my brothers, their English is really, really poor, and my oldest brother, I did ask him for my maths help, but he's terrible at his maths" (as cited in McDaid et al., 2021, p. 36).

Children and Young People with Disability

For children and young people with specific learning needs or a disability, disruption to schooling and respite care placements have the potential to push some families into crisis. Limited access to health services, such as rehabilitation, potentially increases functional limitations for those with chronic conditions. A survey by *Children and Young People with a Disability* revealed most respondents were self-isolating prior to school closures and were unable to buy essentials (Bessell, 2022, p. 5). When students with a disability stayed home from school, this could also send families further into poverty:

> Her son [who has a disability] does attend school and is provided with significant support. School is the time when she has respite and works casually. She has lost her employment and is now unable to afford some of the additional services she was receiving and her son too, such as from an occupational therapist and psychologist. (Advocacy Groups and Services, as cited in Brown et al., 2020, p. 15)

The presence of a sibling with a disability in the home may impact parents/carers' abilities to meet the new demands of home schooling for other children and to manage family stress. Moreover, during lockdowns families no longer had access to respite carers coming into their home, increasing the pressures they were under (Brown et al., 2020). Participation in speech and occupational therapy and other supports provided within school settings led to some children missing basic supports. A disability advocate in our research stated that:

> …a lot of people pulled their children out of school, and then a lot of people decided not to have any other supports coming into the home. Now, that's problematic, because education and learning is not just specifically meeting a particular competency standard of the curriculum, it's also about learning all these skills that able bodied people probably don't understand. So having the access to additional speech therapies or implementing programs that can help children build their skills and their capacity to do everyday living tasks. From that perspective, that was problematic because they weren't having that support, in order for them to be able to do the tasks that they needed to do. Some families stopped professionals coming in or even support workers that may have helped individuals in the school environment.

The Australian state of Victoria had extensive lockdowns throughout 2020, with more than 100 days of school closures. It is therefore unsurprisingly that the number of student expulsions from schools dropped considerably, from 184 in 2019 to 68 in 2020. Disturbingly, however, the proportion of the students with a disability who were expelled rose from one in seven to almost one in three (Carey, 2022). The disruptions to school routines and supports, and the additional stresses both on students with disability and on school staff, thus created yet another level of inequity and marginalisation for these students.

The easing of lockdown restrictions around the world in 2022 created new difficulties for people with disabilities who may have co-morbidities putting them at higher risk of serious illness if infected with COVID-19. With infection rates rising in 2022, students with disabilities may face continued personal isolation, while their peers enjoy renewed freedoms and access to learning back at school.

Poverty

Globally, the COVID-19 pandemic and associated lockdowns have exacerbated pre-existing problems of poverty and precarity. Studies of economic crises as well as birth cohort studies in high as well as low- and middle-income countries demonstrate the potential for life-long negative effects of exposure to poverty early in life, with more severe effects associated with longer exposure to poverty (Yoshikawa et al., 2020, p. 189).

The pandemic has contributed to a significant economic downturn placing additional families in financial distress. In response, the Australian government legislated new, time limited, social safety nets such as "Job Keeper" in April 2020, which ended during March 2021. "Job Seeker", a coronavirus financial supplement for eligible income support recipients, was provided from late April 2020 until late March 2021 (ABS, 2021). During outbreaks in 2021 the Australian government provided income support to support workers to stay at home at a reduced level of pay and with additional eligibility stipulations.

Australian households are experiencing heightened economic vulnerability during the COVID-19 pandemic, and it has been reported that half of Australians (49%) have drawn on unsustainable, finite resources (such as credit or superannuation) to manage their household expenses, which is likely to reduce the amount of money available to be spent on food (Kent et al., 2020, p. 2). Experiencing just a single year of poverty during childhood is associated with poorer outcomes in terms of educational attainment, labour market performance and even overall life satisfaction in early adulthood (Vera-Toscano & Wilkins, 2020, p. 69).

Globally, the pandemic has revealed a lack of effective national policies to tackle poverty, particularly child poverty (Eurochild, 2020). In Australia, prior to the pandemic, almost 18% of Australian children lived in poverty (Davidson et al., 2020). Bessell (2022, p. 4) observes that the COVID-19 pandemic has brought contradictory impacts regarding child poverty in Australia because some 424,000 Australians, including 65,000 children under the age of 15 were suddenly lifted out of poverty through "Job

Seeker" payments in 2020[1]. Among Job Keeper recipients[2], over 3 in 5 (64%) were receiving less income than their usual pay (Bessell, 2022, p. 4). Many children in Australia for the first time were going hungry as food-banks scrambled to reconfigure services to meet the rising tide of food insecurity (Sinha et al., 2020). Children were no longer able to access free food at school, such as through school breakfast clubs, although some schools organised food packs to be dropped off at students' homes (MacDonald, 2020).

Families experiencing homelessness and at risk of homelessness are among Australia's most socially and economically disadvantaged. In recognition that those experiencing homelessness (or at risk of becoming homeless) would be affected by the crisis, state and territory governments adopted a variety of policy responses aimed at boosting housing security such as rent freezes and stays on evictions. During and following COVID-19 lockdowns, people experiencing homelessness also experienced disruptions to basic services for food, personal hygiene, healthcare, and emotional support, and had the added fear of compromised safety. A Specialist Housing Support service worker reported that:

> No school means these kids are not being seen that probably need to be seen. No face-to-face. So only if they're under some sort of medical care are these kids getting seen. So, I think there is going to be an increase of some pretty nasty stuff. (as cited in Robinson, 2020, p. 11)

In Australia, couch surfing and other informal supports that young people access for shelter were affected when public health measures capped the number of extra people who could be present in a home. For some young people, couch surfing occurred against a backdrop of actual or potential fines for non-compliance with physical distancing regulations or households having too many visitors present (Walsh, 2020). Clearly, maintaining an engagement with learning in this context is particularly fraught. One worker described it in this way:

> Like, you know, they could log on and do the online stuff if they wanted to. But none of them have computers, none of them have the internet. But

[1] An Australian federal government welfare incentive for individuals who lost their employment because of lockdowns.

[2] An Australian federal government welfare incentive to help businesses retain staff during lockdowns.

> yeah, they [schools] did give them work packages, like a little workbook to do. When I picked up the work from [school], [I said], 'Oh sorry, you got a pencil or something to give these kids?'...These kids are couch surfing! They don't have a pencil and paper and colouring pencils in their bloody back pockets. (as cited in Robinson, 2020, p. 9)

Precarity

During the COVID-19 pandemic, to some extent all children and young people experienced vulnerabilities in the sense of having their sense of safety reshaped. The final part of this chapter draws together ideas about gender and identity, racism and discrimination, and poverty to focus on a new precariat that appeared due to the indirect effects of COVID-19. The policy response to the new precariat underscored narratives that exists in many industrialised nations between the "deserving" and "undeserving" poor (Katz, 2013). The "deserving" poor are depicted as worthy of financial assistance from the government because they are poor through no fault of their own—such as farmers who are affected by drought. While the "undeserving" poor, in this narrative, are seen as to blame for their situation, such as those exiting prison and in need of housing or other financial help. These categories are often racialised and gendered.

Standing's (2012, 2014) term "precariat"[3] offers a useful frame in the context of COVID-19 and Australia. As discussed earlier in the chapter, during the pandemic Australian household incomes declined, the precariat lived on the edge of unsustainable debt and in chronic economic uncertainty. According to Standing (2014) these defining dimensions produce a consciousness of relative deprivation and a combination of anxiety, anomie (despair of escape from their precarious status), alienation (having to do what they do not wish to do while being unable to do what they would like to do and can do), and anger. In the context of COVID-19 many Australian families experienced vulnerability for the first time and, we argue, joined the precariat.

Australian data suggests that many Australians felt lonely and overwhelmed as the effects of the pandemic unfolded. Conflicts between work and family life generated by school attendance restrictions and the increased pressure of keeping work hours while caring for children likely

[3]For a critique of the term see: Munck, R. (2013). The Precariat: a view from the South. *Third World Quarterly*, *34*(5), 747–762.

contributed to a rise in the mental distress of employed parents/carers (Broadway et al., 2020, p. 7). In the first year of the pandemic 1 in 4 Australians felt like everything was an effort (April 2020, ABS, 2020) and 19%, experienced loneliness (October 2020, ABS, 2020). As the pandemic has drawn on many people have found ways of navigating their loneliness, with the proportion experiencing loneliness dropping to 10% in April 2021 (ABS, 2020). In 2020 COVID-19 affected almost all countries with more than 50 million people around the world acquiring the virus, and more than half of the world's population had experienced a lockdown with strong containment measures (OECD, 2020). Lockdowns eased in 2022, and although loneliness is likely to be affecting fewer people, they may be affected more deeply because some of the most vulnerable people in society (including children) need to continue isolation practices to safeguard their health or no longer have access to leisure activities due to loss of income.

Many parents and caregivers in paid employment had to exhaust their leave entitlements to support learning at home and/or juggle care and work responsibilities without the safety net of support provided by grandparents or extended family. Gaps in policy considerations placed increased stress on families and they struggled to balance workforce and caring responsibilities. The impacts were particularly profound for workers who are in casual employment, where wage earners lost their job, where caregivers needed to continue to work full time, or where they needed to support learning for children in the early years. Such rapid changes can have flow-on effects to the family's finances and stress levels, as well as a general increase in the responsibilities of parents/carers (AIHW, 2021b, p. 11). Capacity to nurture and support learning was stretched. Over half (59%) of parents/carers working from home said their children were having difficulties concentrating while learning from home (AIHW, 2021b).

With increased workloads, more fathers than had previously been the case reported high levels of subjective time stress, unfairness, and dissatisfaction with actively juggling (too much) paid work and unpaid domestic labour and care (Craig & Churchill, 2021, p. 77). An Advocacy Groups and Services senior manager in an NGO that supports marginalised families commented that: "For parents it was too hard to engage children in learning. Parents found it hard to juggle work and school. One staff member said 90% of parents and children didn't enjoy it". Another education system officer in our study explained:

> We have a number of families that were actually doing the learning component when they got home from their own jobs. And so that student's learning day might start later in the day when a parent was there to supervise them, and they may have had a grandparent with them earlier.

We regard families with children in the early years as experiencing vulnerability in the context of COVID-19 due to the higher level of supervision and care they need compared to older children. Parent-only childcare more than doubled compared to pre-pandemic rates, reaching 64% during restrictions (AIFS, 2020). One early childhood education and care sector informant described a multilayered challenge:

> I suppose one of the biggest challenges particularly with the families that have children within that [early years] cohort, is that everything became online. And that's very challenging when you've got students that need hands on learning. So really, where they say, the parents didn't have to be the expert, in a way for these children, these families, they sort of had to. And there are certain levels of capacity, depending on the family, some are very much on top of it all and others aren't. So they struggled a lot in how best to engage. And I think it got to a point where there wasn't that same engagement with school that they would have done, but more trying to keep, you know, the children occupied, and maybe a lot more screen time and less routine than they would have otherwise had with on-site learning.

Providing quality early years education is particularly difficult over an extended period within the home yet is critically important to lifelong education, health, and wellbeing outcomes. The Australian government introduced a free early childhood education and care package from July 2020, but this was time limited.

It was not only parents/carers of young children whose experiences were exacerbated by precarity. Many migrants or refugee students in Australia lost their jobs because of COVID-19 (Mupenzi et al., 2020). This loss of employment further aggravated their socioeconomic situation, worsening the myriad challenges these families already experience and increasing financial hardship. The responsibility to provide remittances for family members is particularly acute for resettled refugees (Mupenzi et al., 2020). Financial precarity was already likely to affect access to the technologies that could support learning at home, such as the ability to buy devices with microphones and camera or have access to a stable high-speed internet, all of which can change the quality of student engagement and impact access to learning material. Many young people and families lack guidance and information about available services and the types of help they are eligible for which is particularly problematic in a period of widespread confinement (Haleemunnissa et al., 2021).

Conclusion

This chapter has considered broad structural issues that influence inequality and contribute to educational vulnerability. Despite the large amount of information collected during the pandemic, significant data gaps persist, particularly for gender-disaggregated data (Flor et al., 2022). The chapter described how the impacts of the COVID-19 pandemic on education are disproportionately born by marginalised and disadvantaged groups and intersect with questions of economic justice. In the next chapter we turn from considering the political economy of educational vulnerability to focus on lessons from past and current crises.

In Australia it has been estimated that one in four children would need specific help to regain learning lost during 2020 (Sonnemann & Goss, 2020), and up to 46% were at risk of adverse effects on their educational outcomes, nutrition, physical movement, social and emotional wellbeing by being physically disconnected from school (Brown et al., 2020) The chapter argues that the experience of vulnerability and precarity *is a consequence* of policy choices. Developing comprehensive strategies that consider structural issues such as child poverty, racism and discrimination is necessary. Making explicit the decisions and stances that work against improving educational outcomes for those who are marginalised is part of countering and advocating for strategies that emphasise a more inclusive and fair future.

References

Australian Bureau of Statistics. (2012). *Census of Population and Housing: Characteristics of Aboriginal and Torres Strait Islander Australians, 2011.* Retrieved June 11, 2021, from https://www.abs.gov.au/ausstats/abs@.nsf/lookup/2076.0main+features702011

Australian Bureau of Statistics. (2020). *Household impacts of COVID-19 Survey, 29 April - 4 May 2020.* Retrieved June 11, 2021, from https://www.abs.gov.au/statistics/people/people-and-communities/household-impacts-covid-19-survey/29-apr-4-may-2020

Australian Bureau of Statistics. (2021). *One year of COVID-19: Aussie jobs, business and the economy.* Retrieved June 11, 2021, from https://www.abs.gov.au/articles/one-year-covid-19-aussie-jobs-business-and-economy

Australian Institute of Family Studies. (2014). *The longitudinal study of Australian Children Annual Statistical Report 2013.* Retrieved June 11, 2021, from https://growingupinaustralia.gov.au/sites/default/files/asr2013.pdf

Australian Institute of Family Studies. (2020). *Families in Australia survey: Life during Covid-19.* Report no. 1: Early findings. https://aifs.gov.au/sites/default/files/publication-documents/covid-19-survey-report_1_early_findings_0.pdf

Australian Institute of Health and Welfare. (2021a). *Youth detention population in Australia 2020* (Cat. no. JUV 135). Retrieved June 11, 2021, from https://www.aihw.gov.au/reports/youth-justice/youth-detention-population-in-australia-2020/contents/summary

Australian Institute of Health and Welfare. (2021b). *Child protection in the time of COVID-19* (Cat. no. CWS 76). Retrieved June 11, 2021, from https://www.aihw.gov.au/reports/child-protection/child-protection-in-the-time-of-covid-19/summary

Australian Institute of Health and Welfare. (2021c). *Data tables: Child protection Australia 2017–18. Child Welfare series no. 70. Cat no. CWS 65.* Australian Institute of Health and Welfare.

Azevedo, J. P., Hasan, A., Goldemberg, D., Iqbal, S. A., & Geven, K. (2020, June 18). Simulating the potential impacts of COVID-19 school closures on schooling and learning outcomes: A set of global estimates. *World Bank Group: Education Global Practice.* https://www.worldbank.org/en/topic/education/publication/simulating-potential-impacts-of-covid-19-school-closures-learning-outcomes-a-set-of-global-estimates

Bandura, A., & Walters, R. H. (1977). *Social learning theory.* Prentice Hall.

Bessell, S. (2022). The impacts of COVID-19 on children in Australia: Deepening poverty and inequality. *Children's Geographies, 20*(4), 448–458. https://doi.org/10.1080/14733285.2021.1902943

Bradbury-Jones, C., & Isham, L. (2020). The pandemic paradox: The consequences of COVID-19 on domestic violence. *Journal of Clinical Nursing, 29*(13–14), 2047–2049. https://doi.org/10.1111/jocn.15296

Broadway, B., Méndez, S., & Moschion, J. (2020). *Behind closed doors: The surge in mental distress of parents.* Melbourne Institute of Applied Economic and Social Research, The University of Melbourne. https://findanexpert.unimelb.edu.au/scholarlywork/1486519-behind-closed-doors%2D%2Dthe-surge-in-mental-distress-of-parents

Bronfenbrenner, U. (1977). Toward an experimental ecology of human development. *The American Psychologist, 32*(7), 513–531. https://doi.org/10.1037/0003-066X.32.7.513

Bronfenbrenner, U. (1995). The bioecological model from a life course perspective: Reflections of a participant observer. In P. Moen, K. Lüscher, & G. H. Elder Jr. (Eds.), *Examining lives in context: Perspectives on the ecology of human development* (pp. 599–618). American Psychological Association.

Brown, N., te Riele, K., Shelley, B., & Woodroffe, J. (2020). *Learning at home during COVID-19: Effects on vulnerable young Australians* (Independent Rapid Response Report). Peter Underwood Centre for Educational Attainment, University of Tasmania. https://www.utas.edu.au/__data/assets/pdf_file/0008/1324268/Learning-at-home-during-COVID-19-updated.pdf

Burgess, C., Bishop, M., & Lowe, K. (2022). Decolonising Indigenous education: The case for cultural mentoring in supporting Indigenous knowledge reproduction. *Discourse: Studies in the Cultural Politics of Education, 43*(1), 1–14.

Butler, J. (2016). Rethinking vulnerability and resistance. In J. Butler, Z. Gambetti, & L. Sabsay (Eds.), *Vulnerability in resistance* (pp. 12–27). Duke University Press.

Carey, A. (2022, April 12). 'Devastating impact': Rise in proportion of disabled students expelled during pandemic. *The Age* (online). https://www.theage.com.au/politics/victoria/devastating-impact-rise-in-proportion-of-disabled-students-expelled-during-pandemic-20220403-p5aad3.html

Chakraborty, D. (2020). The "living dead" within "death-worlds": Gender crisis and covid-19 in India. *Gender, Work & Organization, 28*(S2), 330–339. https://doi.org/10.1111/gwao.12585

Cohen, J., & van der Meulen Rodgers, Y. (2021). The feminist political economy of Covid-19: Capitalism, women, and work. *Global Public Health, 16*(8–9), 1381–1395. https://doi.org/10.1080/17441692.2021.1920044

Craig, L., & Churchill, B. (2021). Dual-earner parent couples' work and care during COVID-19. *Gender, Work & Organization, 28*(S1), 66–79. https://doi.org/10.1111/gwao.12497

Cross-Townsend, M. (2011). Indigenous education and Indigenous studies in the Australian academy: Assimilationism, critical pedagogy, dominant culture learners, and Indigenous knowledges. *Counterpoints, 379*, 68–79.

Davidson, P., Saunders, P., Bradbury, B., & Wong, M. (2020). *Poverty in Australia 2020: Part 1, Overview*. ACOSS/UNSW Poverty and Inequality Partnership Report No. 3. ACOSS.

Devakumar, D., Shannon, G., Bhopal, S. S., & Abubakar, I. (2020). Racism and discrimination in COVID-19 responses. *The Lancet, 395*(10231), 1194. https://doi.org/10.1016/S0140-6736(20)30792-3

Dixson, A. D., & Rousseau Anderson, C. (2018). Where are we? Critical race theory in education 20 years later. *Peabody Journal of Education, 93*(1), 121–131.

Eurochild. (2020). *Growing up in lockdown: Europe's children in the age of COVID-19*. Eurochild. Retrieved December 13, 2020, from https://eurochild.org/uploads/2020/12/2020-Eurochild-Semester-Report.pdf

Fabrizio, S., Gomes, D. B. P., & Tavares, M. M. (2021). *Covid she-cession: The employment penalty of taking care of young children* (IMF Working Papers Volume 2021, Issue 058). International Monetary Fund. https://doi.org/10.5089/9781513571157.001

Fish, J. N., McInroy, L. B., Paceley, M. S., Williams, N. D., Henderson, S., Levine, D. S., & Edsall, R. N. (2020). "I'm kinda stuck at home with unsupportive parents right now": LGBTQ youths' experiences with COVID-19 and the importance of online support. *Journal of Adolescent Health, 67*(3), 450–452. https://doi.org/10.1016/j.jadohealth.2020.06.002

Flor, L. S., Friedman, J., Spencer, C. N., Cagney, J., Arrieta, A., Herbert, M. E., Stein, C., Mullany, E. C., Hon, J., Patwardhan, V., Barber, R. M., Collins, J. K., Hay, S. I., Lim, S. S., Lozano, R., Mokdad, A. H., Murray, C. J. L., Reiner, R. C., Jr., & Gakidou, E. (2022). Quantifying the effects of the COVID-19 pandemic on gender equality on health, social, and economic indicators: A comprehensive review of data from March, 2020, to September, 2021. *The Lancet, 399*(10344), 2381–2397. https://doi.org/10.1016/S0140-6736(22)00008-3

Foley, M., & Williamson, S. (2021). Women, work and industrial relations in Australia in 2020. *Journal of Industrial Relations, 63*(3), 321–337. https://doi.org/10.1177/0022185621996407

Griffiths, K., Coleman, C., Lee, V., & Madden, R. (2016). How colonisation determines social justice and Indigenous health—A review of the literature. *Journal of Population Research, 33*(1), 9–30. https://doi.org/10.1007/s12546-016-9164-1

Haleemunnissa, S., Didel, S., Swami, M. K., Singh, K., & Vyas, V. (2021). Children and COVID19: Understanding impact on the growth trajectory of an evolving generation. *Children and Youth Services Review, 120*, 105754. https://doi.org/10.1016/j.childyouth.2020.105754

Hill Collins, P., & Bilge, S. (2020). *Intersectionality* (2nd ed.). Wiley.

International Labour Organization, & United Nations Children's Fund. (2020). *COVID-19 and child labour: A time of crisis, a time to act.* https://www.ilo.org/wcmsp5/groups/public/%2D%2D-ed_norm/%2D%2D-ipec/documents/publication/wcms_747421.pdf

Jani, A. (2020). Preparing for COVID-19's aftermath: Simple steps to address social determinants of health. *Journal of the Royal Society of Medicine, 113*(6), 205–207. https://doi.org/10.1177/0141076820921655

Katz, M. (2013). *The undeserving poor: America's enduring confrontation with poverty* (2nd ed.). Oxford University Press.

Kent, K., Murray, S., Penrose, B., Auckland, S., Visentin, D., Godrich, S., & Lester, E. (2020). Prevalence and socio-demographic predictors of food insecurity in Australia during the COVID-19 pandemic. *Nutrients, 12*(9), 2682. https://doi.org/10.3390/nu12092682

Levitas, R., Pantazis, C., Fahmy, E., Gordon, D., Lloyd-Reichling, E., & Patsios, D. (2007). *The multi-dimensional analysis of social exclusion*. University of Bristol.

MacDonald, F. (2020). Schools provide food for many hungry children. This needs to continue when classes go online. *The Conversation*. Retrieved January 30, 2021, from https://theconversation.com/schools-provide-food-for-many-hungry-children-this-needs-to-continue-when-classes-go-online-134384

Marmot, M., & Wilkinson, R. (Eds.). (2005). *Social determinants of health* (2nd ed.). Oxford University Press.

McDald, L., Povey, J., Tomaszewski, W., Cleary, A., Plage, S., Cook, S., Xiang, N., Zajac, T., Ridgway, T., & Western, M. (2021). *What do children, young people and families say about needs, and impact of COVID-19?* (Pillar Report No. 2). Institute for Social Science Research, University of Queensland. https://issr.uq.edu.au/files/21372/Ramsay%20Pillar%202_Final.pdf

Munck, R. (2013). The Precariat: a view from the South. *Third World Quarterly, 34*(5), 747–762.

Mupenzi, A., Mude, W., & Baker, S. (2020). Reflections on COVID-19 and impacts on equitable participation: The case of culturally and linguistically diverse migrant and/or refugee (CALDM/R) students in Australian higher education. *Higher Education Research & Development, 39*(7), 1337–1341. https://doi.org/10.1080/07294360.2020.1824991

OECD. (2020). *The territorial impact of COVID-19: Managing the crisis across levels of government*. https://www.oecd.org/coronavirus/policy-responses/the-territorial-impact-of-covid-19-managing-the-crisis-across-levels-of-government-d3e314e1/

Poteat, T., Millett, G. A., Nelson, L. E., & Beyrer, C. (2020). Understanding COVID-19 risks and vulnerabilities among black communities in America: The lethal force of syndemics. *Annals of Epidemiology, 47*, 1–3. https://doi.org/10.1016/j.annepidem.2020.05.004

Productivity Commission. (2020). *Overcoming Indigenous Disadvantage Key Indicators 2020* (Fact Sheet No. 1). https://www.pc.gov.au/research/ongoing/overcoming-indigenous-disadvantage/2020/report-documents/oid-2020-factsheet.pdf

Robinson, C. (2020). *#StayHome? The impact of COVID-19 on unaccompanied homeless children in Tasmania*. .

Salerno, J. P., Williams, N. D., & Gattamorta, K. A. (2020). LGBTQ populations: Psychologically vulnerable communities in the COVID-19 pandemic. *Psychological Trauma: Theory, Research, Practice, and Policy, 12*(S1), S239–S242.

Sharma, S. V., Chuang, R. J., Rushing, M., Naylor, B., Ranjit, N., Pomeroy, M., & Markham, C. (2020). Peer reviewed: Social determinants of health–related needs during COVID-19 among low-income households with children.

Preventing Chronic Disease, 17, 200322. https://doi.org/10.5888/pcd17.200322
Sinha, I., Bennett, D., & Taylor-Robinson, D. C. (2020). Children are being sidelined by covid-19. *BMJ, 369*. https://doi.org/10.1136/bmj.m2061
Sonnemann, J., & Goss, P. (2020). *COVID catch-up: Helping disadvantaged students close the equity gap* (Grattan Institute Report No. 2020-08, June 2020). Grattan Institute, Victoria. https://grattan.edu.au/wp-content/uploads/2020/06/COVID-Catch-up-Grattan-School-Education-Report.pdf
Standing, G. (2012). The precariat: From denizens to citizens? *Polity, 44*(4), 588–608. https://doi.org/10.1057/pol.2012.15
Standing, G. (2014). The precariat. *Contexts, 13*(4), 10–12. https://doi.org/10.1177/1536504214558209
United Nations. (1989). *Convention on the Rights of the Child*. General Assembly Resolution 44/25, UN Doc. A/RES/44/25. UN.
United Nations Entity for Gender Equality and the Empowerment of Women. (2020). *COVID-19 and ending violence against women and girls*. https://www.unwomen.org/en/digital-library/publications/2020/04/issue-brief-covid-19-and-ending-violence-against-women-and-girls
Vera-Toscano, E., & Wilkins, R. (2020). *Does poverty in childhood beget poverty in adulthood in Australia?* (Report Series No. 1). Melbourne Institute: Applied Economic & Social Research, University of Melbourne.
Walker, R., Usher, K., Jackson, D., Reid, C., Hopkins, K., Shepherd, C., Smallwood, R., & Marriott, R. (2021). Connection to… Addressing digital inequities in supporting the well-being of young Indigenous Australians in the wake of COVID-19. *International Journal of Environmental Research and Public Health, 18*(4), 2141. https://doi.org/10.3390/ijerph18042141
Walsh, K. (2020, August 3). *If not now, when?* Pro Bono Australia. https://probonoaustralia.com.au/news/2020/08/if-not-now-when/
Yoshikawa, H., Wuermli, A., Britto, P. R., Dreyer, B., Leckman, J. F., Lye, S. J., Ponguta, L. A., Richter, L. M., & Stein, A. (2020). Effects of the global coronavirus disease-2019 pandemic on early childhood development: Short- and long-term risks and mitigating program and policy actions. *The Journal of Pediatrics, 223*, 188–193. https://doi.org/10.1016/j.jpeds.2020.05.020

CHAPTER 3

Lessons from Crises and Disasters: Then and Now

Abstract This chapter reviews national and international literature to look specifically at significant crises in the past one hundred years, focussing mainly on developed countries, and how such crises have affected how children and young people access education. The selected crises include influenza pandemics (1918–19 "flu pandemic"; H1N1/09 "swine flu"), the COVID-19 pandemic, and environmental disasters including Hurricane Katrina, United States (2005), Christchurch Earthquakes, New Zealand (2010–11), Great East Japan Earthquake (2011), and bushfires in Australia (2009, 2019/20).

This chapter examines three questions: How has school education responded to crises in the past? How did these crises and responses impact on learners? And what lessons can be learned about how crises affect social and educational equity? We answer these in three sections. Section 1 introduces and contextualises the selected historical crises. Section 2 identifies three key impacts of crises on learners: first, that access to schooling is compromised; second, that school closures impact on student mental health and wellbeing; and third, there is an increased risk of academic under-achievement. Section 3 draws out the key lessons that can be learned from these crises which are useful in informing how education systems prepare for managing crises into the future.

Keywords Crisis • Disaster • Education • Influenza pandemic • H1N1/09 • Swine flu pandemic • COVID-19 pandemic • Hurricane

E. S. Rudling et al., *Education and Equity in Times of Crisis*,
https://doi.org/10.1007/978-3-031-18671-4_3

Katrina • Christchurch Earthquakes • Great East Japan Earthquake • Bushfires in Australia • Recovery • Mental health • Emotional wellbeing • Learning loss • Displacement

Introduction

Over the past one hundred years, various crises have precipitated school closures resulting in the displacement of students, yet the impact of crises on student engagement with learning with and academic achievement is poorly understood. This, in part, is due to the demanding nature of a crisis: the focus is on immediate response, survival and recovery. However, given the importance of schooling to students' emotional and mental wellbeing and lifelong productivity, there are lessons to be learned from past crises to understand how to respond to future similar events with a focus on safeguarding equity in education.

Disaster and crisis are sudden and disruptive events. Often life-threatening, they can impact large areas or be localised to a particular community. Smith and Riley (2012, p. 53) describe crisis as a "confronting, intrusive and painful experience" that requires immediate and decisive action. In this chapter, we highlight five crises that can be broadly distinguished into two categories: viral crises and environmental disasters.

As we will discuss, such events impact on schools and education by disconnecting students from their regular place of learning, as well as from the routines of learning, and the social and emotional support provided by educators and their peers. Additionally, in such emergencies, schools often transform from places of learning into community disaster management hubs yet the impact of this complex crisis scene on students' access to education is under researched.

Although crisis affects people differently depending on their socio-economic and educational context (see Chap. 2), children and young people are identified as being particularly vulnerable to the effects of crisis and disaster (Davis et al., 2021; Peek, 2008; Peek et al., 2017):

> Children are psychologically vulnerable and may develop post-traumatic stress disorder or related symptoms; are physically vulnerable to death, injury, illness, and abuse; and often experience disruptions or delays in their educational progress as a result of disaster. (Peek, 2008, p. 1)

Beyond the immediate concerns pertaining to safety, mental health, and wellbeing, throughout and beyond crisis students are also vulnerable to learning loss and may disengage altogether from learning (McDaid et al., 2020) which has longer term implications for life satisfaction and overall productivity (Azevedo et al., 2020; Brown et al., 2020; McDaid et al., 2020).

To articulate the necessity of understanding how crisis and disaster affect children and young people and their educational attainment, this chapter highlights salient lessons from historical viral crises and environmental disasters focusing on how they affected children and young people and their engagement with learning and education. We introduce the selected disasters and crises in Section 1. We acknowledge there have been many more crises around the globe in the past century, including due to war and armed conflict, with negative consequences for children's education. However, we focus on the selected crises due to the availability of relevant research and the relative comparability of school systems of the affected nations.

Common to the chosen events is that they impacted on students' ability to access school and participate in their education. In Section 2, we conduct a detailed analysis of what happens to education during a crisis. We probe into the nature of crises by drawing together research findings from past events and from the present COVID-19 pandemic to elicit insights into the impacts of crises on learners and education systems.

Emerging from a crisis is not a clear or simple process (Davis et al., 2021, p. 1). In Section 3, we address broad lessons learned from recovery from historical crises to identify what students need in order to continue with their education throughout and beyond disaster. When considered as a collective; past and present crises offer insight into the key challenges and opportunities in safeguarding equitable access to education for children and young people during, and following, times of crisis and disaster.

Disaster and crisis often expose, if not worsen, pre-existing educational inequities. This can be illustrated by the inequitable impact of COVID-19 induced school closures on student learning, as discussed throughout this book. Cahill et al. (2020a) indicate that the coronavirus pandemic "put additional demands on families…those without sufficient resources or support at home thus faced additional inequity" (p. 11). It is imperative that we learn from historical crises to address educational equity now and in future challenges (Rees et al., 2021).

Section 1: Introducing the Selected Crises

Here, we introduce the five selected crises and disasters that foreground our discussion of the impact of such emergencies on school students. Cauchemez et al. (2014) distinguish between reactive school closures—in response to outbreaks or disaster—and proactive school closures—implemented prior to an outbreak of disease (p. 2). While we recognise the value in this, we have distinguished the crises into two broad categories based on type: viral health crises and environmental disasters, as it is type that tends to dictate the resulting school closure.

Viral Health Crises

Over the past one hundred years, air-borne viruses, including influenza pandemics and the COVID-19 pandemic, have forced the closure of schools to reduce the severity of the outbreak (Garbe et al., 2020; Viner et al., 2020). Two notable instances of influenza induced school closures include the 1918–19 influenza pandemic, and the 2009 H1N1/09 "swine flu" crisis, which together set the precedent for using school closures to manage later virus outbreaks like the coronavirus pandemic (Haug et al., 2020). Common to both the 1918 influenza pandemic and the 2009 H1N1 pandemic, as well as the 2019 coronavirus pandemic, is the implementation of school closures as a non-pharmaceutical intervention to reduce rates of community transmission and infection (Armitage & Nellums, 2020; Haug et al., 2020).

Influenza Pandemic (1918–19)

The influenza pandemic of 1918–19 emerged in the northern spring of 1918 in Midwestern United States and, later that same year, subsequent waves of infection spread globally (Chowell et al., 2006). This was the world's most deadly recent pandemic, with an estimated loss of between 20 and 100 million lives (Chowell et al., 2006). Public health records of the 1918–19 flu pandemic reveal school closures were implemented around the world to mitigate rates of disease transmission. These were generally in effect for relatively short periods of time, although closures lasted up to four months in the United States (Ager et al., 2020; Battenfeld, 2020; Caley et al., 2008).

H1N1 Swine Flu (2009)

In mid-2009, the H1N1 "swine flu" outbreak prompted school closures in numerous countries including Hong Kong, Serbia, France, the United States, Japan, New Zealand, the United Kingdom, and Australia. Like the 1918–19 influenza pandemic, schools were identified as sites with high rates of infection and children were recognised as key transmitters of the virus (Cauchemez et al., 2014). As a result, schools were closed. Globally, the length of school closures varied (Cauchemez et al., 2014); however, in the United States, more than 700 schools closed for one or two weeks during the outbreak (Klaiman et al., 2011).

COVID-19 Pandemic (2019-Ongoing)

The global coronavirus pandemic (commencing December 2019) is widely accepted as the most significant disruption to education recorded (Azevedo et al., 2020; United Nations, 2020; Viner et al., 2020). Worldwide, school closures were implemented to assist with the reduction of community transmission and infection (Armitage & Nellums, 2020; Haug et al., 2020). Monitoring by United Nations Educational, Scientific and Cultural Organization (UNESCO) (2021) shows how the impact of the pandemic on school closures has changed over time. On 17 February 2020, there was 1 country-wide closure (Mongolia) affecting 883,055 learners, or 0.1% of all enrolled learners globally[1]. There were also shifts to home learning in China[2]. By April 2020, there were 151 country wide school closures affecting 81.6% of enrolled learners (approximately 1.3 billion students) globally. In addition, the USA, Russia, and Australia were among the many countries with partial school closures. A year later, in April 2021, the number of closed schools had dropped significantly, with 8.7% of learners worldwide impacted by 26 country-wide school closures and, in February 2022, six country wide closures impacted 2.8% of learners globally (UNESCO, 2021). In 2021 the OECD estimated 1.5 billion students from 188 countries had been affected by school closures (OECD, 2021, p. 3).

Distinguishing COVID-19 school closures from those enacted in response to the influenza pandemic (1918–1919) and the H1N1 pandemic (2009) was that, initially, children and young people had some of the lowest rates of infection of COVID-19 yet were the most adversely impacted by school closures (United Nations, 2020). Mutations of the original *alpha* strain of the coronavirus emerged over during 2021 and 2022 with the *delta* variant then becoming dominant, ahead of the *omicron* variant and sub-variants. While lockdowns and school closures were

[1] There were also partial school closures in other countries, like Viet Nam.

[2] China, the world's most populous nation with more than 1.4 billion people, adopted stringent lockdown policies over multiple years of the COVID-19 pandemic. While this text does not focus on the experiences of learning at home in China, research by Wang et al. (2020) and Zhang et al. (2020) investigates the impacts of home confinement on students in China during the pandemic.

widespread internationally with the *alpha* strain, they typically became more localized with the subsequent variants of the virus, particularly as vaccination programs rolled out in countries (albeit at varying rates of population coverage) (Howard-Jones et al., 2022; Mathieu et al., 2021; Russell et al., 2022). By 2022, school jurisdictions around Australia moved to a policy that prioritised all schools remaining open with management of outbreaks on a local site-by-site basis.

Disasters

Schools are also affected by environmental disasters. Here, we introduce four selected disasters that have disrupted schooling in various nations in the global north. These events, a hurricane, an earthquake, a tsunami, and bushfires, were localised in specific countries. A key difference between virus related school closures and those resulting from disasters is that the former kind of closure is enforced to manage rate of infection. In comparison, school closures resulting from disaster tend to be sudden and cause extensive destruction of infrastructure, including homes, roads, community facilities, and schools themselves.

Hurricane Katrina, USA (2005)

In late August 2005, just prior to the commencement of the American school year, Hurricane Katrina made first landfall in New Orleans, Louisiana, USA. The Category 5 hurricane caused more than 1,800 deaths across Louisiana, Mississippi, and other southern states. Approximately 80% of New Orleans was inundated with catastrophic flooding for weeks and more than 600,000 households were still displaced one month after the hurricane (Plyer, 2016). Fothergill and Peek (2006) estimate Hurricane Katrina displaced up to 300,000 school-aged children, and Kousky (2016) suggests almost 200,000 public school students were required to relocate to other areas or schools to continue with their education.

Christchurch Earthquakes, New Zealand (2010–11)

Between September 2010 and February 2011, multiple damaging earthquakes struck the Canterbury region of New Zealand, and in particular, the city of Christchurch (Clifton et al., 2011; Potter et al., 2015). This culminated with a 6.3 magnitude earthquake hitting Christchurch at midday on 22 February 2011. This catastrophic event injured 6659 people, and 182 deaths were recorded in the first 24 hours of the disaster (Ardagh et al., 2012).

The earthquakes caused substantial destruction to city infrastructure, including to schools (Potter et al., 2015). Three weeks after the February event, 84% of school students in the greater Christchurch area were attending school (Potter et al., 2015). However, the continuation of education required relocating over 12,000 students to new schools, sometimes outside of the Christchurch region, with half of all secondary schools "site sharing" (Potter et al., 2015). Disaster management efforts included the relocation of the staff and students of nine schools to other, undamaged, school sites in the region, a strategy designed to minimise disruption for affected students (Ham et al., 2012).

Great East Japan Earthquake (2011)

On 11 March 2011 a 9.0–9.1 magnitude undersea earthquake in the Pacific Ocean created a tsunami that hit the East Coast of Japan's most populated island, Honshu. The event is known as the Great East Japan Earthquake and is one of the most severe ever recorded. The tsunami struck the Tōhoku prefecture causing extensive damage, including destruction from disastrous flooding. The tsunami compromised the Fukushima Daiichi nuclear power plant in Fukushima Prefecture, triggering the most catastrophic nuclear disaster since Chernobyl in 1986. An estimated 20,000 people were killed or recorded missing and approximately 130,000 houses were levelled, with a projected 270,000 homes severely damaged (Ranghieri & Ishiwatari, 2014). Additionally, 6,284 public schools were compromised by the tsunami, including 193 that were destroyed with a further 747 schools requiring structural repairs (Ranghieri & Ishiwatari, 2014, p. 77). The disaster occurred during school hours with 733 teachers and students killed or declared missing (Ranghieri & Ishiwatari, 2014, p. 77).

Bushfires in Australia (2009, 2019–20)

Over the past two decades, Australia has experienced multiple catastrophic bushfires, causing loss of human and animal life, extensive damage to bushland, and the destruction of buildings and other infrastructure. Bushfires tend to occur in the southern hemisphere summer during the long school holidays (December–February) and can burn for many weeks. When the school year starts (February), student access to education is interrupted by emotional stress and trauma; the destruction of buildings including homes and schools; and loss of property and jobs affecting parental income (Harms et al., 2021).

Two bushfire seasons, 2009 and 2019–2020, were especially devastating to affected communities and schools. In February 2009, the state of Victoria suffered severe bushfires which took 173 lives (Teague et al., 2010)[3]. The summer of 2019/20 saw massive, uncontrolled bushfires burn for months, across large regions in the states of New South Wales (NSW), South Australia (SA) and Victoria. In each of these areas, small towns experienced extensive property damage and numerous lives were lost. The 2019/20 bushfire season damaged 178 schools in NSW alone (Chrysanthos, 2020).

Disaster and crisis impact on students' access to schooling. Each example highlights how disasters and crises affect the education of school students. These selected events form the basis of our thematic analysis of the impacts of crises and disasters on learners, learning, and school systems more broadly—especially in relation to those experiencing vulnerability prior to the disaster.

Section 2: Three Impacts of Disaster and Crisis on the Learner

Here, we focus on some of the immediate impacts of crises on the learner. Simply put, crises are substantive events that disrupt normal life for children and young people by destroying infrastructure, interrupting routines, and impacting on access to services and supports including school. As such, the first key theme we raise is that of enforced relocation to address how disaster and crises separate students from physically attending their school of enrolment. Secondly, we discuss how disaster and crisis negatively impact on student mental health and wellbeing, and that recovery takes time. Third, we explore how these issues, together, contribute to increased likelihood of student learning loss and disengagement from learning.

Our analysis is framed by our theoretical lens introduced in Chap. 1. In considering schooling and education from an ecological system perspective (Bronfenbrenner, 1974, 1977), school closures create interruptions and disturbances across system levels. At the most obvious level, the microsystem, school closures profoundly reorient interactions in and between the home environment and students' everyday settings of the school classroom. Employing an ecological system view (Bronfenbrenner,

[3] The disaster is now known as Black Saturday.

1977, 1995), we locate the learner at the centre of the ecological system, and, in this section, we consider the interruptions brought by crises and their impacts on the learner's interactions in their "everyday environments" of the home and the classroom at the microsystem level.

Access to Schooling Is Compromised

Crisis, be it an environmental catastrophe or a viral health pandemic, interrupts students' access to education. In terms of disasters, extensive destruction of infrastructure means students either temporarily cease schooling or are relocated to another school or temporary place of learning in order to continue with their education (see Potter et al., 2015). For example, the extensive devastation wrought by the Great East Japan Earthquake (2011) included the destruction of almost two hundred schools and many. Students were relocated to new schools in unaffected regions to continue with their learning (Ranghieri & Ishiwatari, 2014). Students were also relocated to new schools following the Christchurch Earthquakes (2010–11) and Hurricane Katrina (2005). In comparison, in response to pandemics like COVID-19, school closures moved learning from the school to the home environment (Brown et al., 2020).

The random nature of disaster means the impact of relocation on schooling is uneven, with some students more affected than others. In a study of ninety-four children affected by the Christchurch earthquakes (2010–11), Freeman et al. (2015) identified a range of relocation experiences and that "most children made multiple moves, some within Christchurch and its surrounds while others moved to other towns and cities either temporarily or permanently" (p. 107).

The influenza (1918–19), swine-flu (2009), and COVID-19 (2019) pandemics all precipitated school closures pushing students to learn in the home environment, raising significantly different challenges to those associated with relocation brought on by disaster. First, as outlined in Chaps. 2 and 5, home is not safe for all students. Secondly, as will be argued in Chap. 6, not all students can access learning resources at home. Third, social isolation while learning from home can negatively affect student mental health and wellbeing for many students (see Brown et al., 2020).

Irrespective of whether it is in response to virus or disaster, relocation can exacerbate disadvantage experienced by children and young people. Beyond the trauma of the crisis itself, relocated students must navigate settling into new schools and forming new peer groups. In some disasters, such as Hurricane Katrina, families living in low-lying low socio-economic areas were disproportionately impacted by the inundation when compared with families living in higher socio-economic areas (see Fothergill & Peek, 2006). However, the transfer of students to temporary schools following Hurricane Katrina (2005) also illustrates how relocation can benefit learning. In the immediate aftermath of Hurricane Katrina the transfer of students from low-achieving schools to higher-achieving schools academically benefitted some low-achieving students (Peek & Richardson, 2010).

In contrast, the COVID-19 pandemic school closures highlight how relocating learning to the home environment (rather than to a different school) can be inequitable: not all students have access to safe, well-resourced home environments that support learning. Learning from home blurs the boundaries between home, school, family, parents/carers, and teachers (See McDaid et al., 2020). In recognition that not all home environments are safe, several nations, like Australia and the United Kingdom, continued learning at school for students who did not have access to a safe learning environment at home or for children of essential workers, such as doctors and nurses (McDaid et al., 2020; Moss et al., 2021). These students were commonly placed in temporary class groups that were not organised around grade or age cohorts. In the United Kingdom and Australia, teachers were largely required to simultaneously teach online and in-class, with students receiving the same lesson irrespective of their learning modality. In-class learning relied on teacher assistants teaching prepared lessons to students while teachers taught online (Brown et al., 2020; Moss et al., 2021).

Focusing on COVID-19 related school closures in Australia, scholarly literature shows that there was increased pressure for parents/carers to assist with their child/ren's learning at home (Garbe et al., 2020). However, not all parents/carers had the skills, resources, or capacity to do so. In a study into learning from home in Australia throughout the COVID-19 pandemic, Brown et al. (2020) found that parents/carers felt challenged and confused about their role as both parents/carers and educators during learning from home (p. 21). Another Australian study, by Beasy et al. (2020), identified that parents/carers found learning at home impractical. This was amplified for parents/carers of children with disability (Dickinson & Yates, 2020). Reports of inappropriate caregiver involvement with online classes also increased pressure on educators (Adams, 2021).

Globally, research indicates that school closures impact negatively on student health and wellbeing (Graham et al., 2017; Kneale et al., 2020; Lamb, 2020; Lee, 2020; Shakespeare-Finch et al., 2020). School closures disconnect students from the social, emotional, and health supports provided by schools (McMahon et al., 2000; United Nations, 2020). Kneale et al. (2020) identify that school closures affect children's health and wellbeing and educational attainment, as well as impact on their parents/carers, and families. In turn, reduced health and wellbeing also affects the school itself including both staff and community. Disruptions like school closures extend beyond the microsystem level of students' home and classroom environments, also affecting meso-level interactions between families and teachers and the broader community (Bronfenbrenner, 1995).

Student mental health and wellbeing is not only impacted by both the disaster and/or crisis but is also affected throughout the recovery from the event. Crisis itself can be a source of trauma for many students, not only influencing mental health and wellbeing in the short term, but also affecting student capacities to engage with learning in the longer term (Gibbs et al., 2019). Disasters, such as Hurricane Katrina, the Christchurch Earthquakes, and the Australian bushfires, show how significant destruction exposes students to multiple losses including family, friends, work, sport, routine and structure, and pets (see Cahill et al., 2020b; Freeman et al., 2015; Kousky, 2016). Such events influence students mentally, emotionally, and developmentally.

When school closures are combined with society-wide lockdowns, such as during the COVID-19 pandemic, there is increased likelihood of adverse mental health outcomes for students (Graham et al., 2017; Lamb, 2020; Lee, 2020; Nicolson et al., 2020; Shakespeare-Finch et al., 2020). This holds true for widespread destruction following a catastrophe. Potter et al. (2015) highlight how the Christchurch Earthquakes (2010–11) destroyed key sites of community interaction, like schools and school sporting facilities and events, which negatively impacted on connectedness and local identity. Further, school closures disrupt the ways that schools and families activate social capital (Bourdieu, 1984); through the social networks that provide additional learning and wellbeing opportunities for students, for example through work and sports activities that extend beyond the school.

Scholars caution that the developmental age of students influences their recovery from crisis and disaster (Freeman et al., 2015; Yoshikawa et al., 2020). However, evidence about the effect of crises on students of different ages varies. Research conducted with elementary and middle school students who were relocated following the Great East Japan Earthquake found that younger students were more vulnerable to the onset of paediatric mental disturbance than older students (Lieber, 2017). Further research on the longer term impacts of crises as reported by Gibbs et al. (2019) suggests that younger students experience delayed effects which can affect their educational outcomes in later years. In contrast, following the Christchurch Earthquakes (2010–11), Ham et al. (2012) found that senior secondary students who had to move schools to continue their education reported increased anxiety about their education and post-school pathways whereas junior students were more likely to report positive impacts of the new schooling arrangements.

In terms of the management of the COVID-19 pandemic, there is concern that the focus on safeguarding physical health from infection has overlooked the importance of mental health and wellbeing, which has resulted in a "shadow pandemic" (Nicolson et al., 2020).

Recovery Takes Time

Supporting mental health and wellbeing is important throughout the recovery phase, well beyond the initial disaster or crisis. Recovery is multifaceted and research highlights various strategies implemented by schools to support student recuperation including art therapy, physical supports, and community building.

The recovery from the Christchurch Earthquakes (2010–11) exemplifies the vital role schools play in fostering effective coping strategies in students following crisis. Research conducted with school staff following the Christchurch Earthquakes found instances of traumatised students and an increased need for providing emotional support to students, including facilitating peer connections (Johnson & Ronan, 2014; Mooney et al., 2021; Mutch, 2015). Mooney et al. (2021) noted the importance of "providing routine, consistency and a sense of safety and security" (p. 24). Art therapy and narrative therapy are examples of how Christchurch schools supported students' mental wellbeing following the earthquakes (Mooney et al., 2021). Similarly, following the Great East Japan Earthquake (2011), art and creative therapies were implemented in affected primary schools to assist with rebuilding resilience and to foster children's participation in community rebuilding (Goto et al., 2022). These types of initiatives can help students process the emotional trauma of a crisis as they re-engage with schooling.

The aftermath of the Victorian bushfires (2009) shows both the time recovery takes and the ongoing need for mental health support in school recovery from disaster. Cahill et al. (2020a) observe that, following the fires, instances of mental health problems in children and young people reduced over time but remained higher than the national average (p. 12). Shakespeare-Finch et al. (2020) caution that ongoing climate disaster, including fire, drought, and flooding, in the areas affected by the 2019–2020 fires may have long-term impacts on mental health and well-being of children and young people.

A potential challenge to disaster response is that children and young people react to crisis differently and there is a need for tailored, student-focused approaches to recovery. Gibbs et al. (2019) identified variations in how the Victorian bushfires impacted children, with some being minimally affected, others recovering over time and some students experiencing delayed impacts which, the authors indicated, were difficult for teachers and parents/carers to address. They explain "over time, parents[sic] and schools may not recognise that delayed impacts arise from the disaster experience, and therefore children may not be offered appropriate support programs" (Gibbs et al., 2019, p. 1408). This may be, in part, due to the mental health of parents/carers also being affected by the bushfires up to five years after the event (Bryant et al., 2018). Gibbs et al. (2019) suggested this could change children's home environments in ways that hindered their capacities to study and learn.

The recognition that recovery takes time is especially significant for supporting younger students. Yoshikawa et al. (2020) argue disasters and crises have negative long-term impacts on young children because their brain architecture is still developing and is therefore especially vulnerable to adversity. Early instability and disruption to a child's education can have lifelong consequences in terms of their development, health and learning (Sandstrom & Huerta, 2013; Yoshikawa et al., 2020). Kousky (2016) explains, "early-life health shocks are associated with fewer years of schooling, reduced economic activity, delayed motor development, lower IQ, more behavioural problems, and lower test scores" (p. 81) and that the impacts of these persist across generations. Reflecting on the 1918–19 influenza pandemic, Kousky (2016) explains that while not all children suffered influenza directly, research indicates that early life shocks include exposure to catastrophic events and instability, which means that even children who do not contract the virus can suffer its impacts.

Increased Risk of Academic Under-Achievement

The combined effects of adverse mental health and relocation within the context of crisis and disaster exacerbate the potential for student learning loss. Mental health concerns and relocation together affect academic achievement beyond the immediate crisis event. School closures disconnect students from their school and schooling routines, as well as from key social and emotional supports normally accessed from school.

Research on school closures during the 1918–19 influenza pandemic reveals that, during the period of learning from home, the total number of hours spent learning was reduced (Rich, 2020). This also occurred during COVID-19-related school closures. Grewenig et al. (2020, p. 2) found learning from home "more than halved students' learning time" in Germany for both high and low achieving students.

However, students do continue to learn—including informally (see Bourke et al., 2021). During the 1918–19 influenza virus school closures, available research indicates there was little focus on completing what schoolwork was sent home as students instead assisted with chores and work around the home (Rich, 2020). Focusing on the COVID-19 pandemic, in Australia, educators have estimated students learned at home for 2–3 hours per day and that "formal learning was being supplemented by an increase in informal learning, such as cooking, gardening, and craft activities facilitated by families" (Brown et al., 2020, p. 28). In Germany, Grewenig et al. (2020) identified that that high achieving students spent more time exercising or following creative pursuits than low achieving students who spent more time watching television, gaming or on social media (pp. 7–8).

Irrespective of the broad benefits of informal learning to student development, research indicates that the reduction in time spent learning has negative effects for educational attainment. The potential for learning loss is illustrated by standardised testing results following the 2009 Victorian bushfires in Australia. Gibbs et al. (2019) compared student learning data in National Assessment Program—Literacy and Numeracy (NAPLAN) scores for children attending high, medium, and low disaster-affected primary schools and found greater reductions in expected learning gains from Year 3 to Year 5 test results in schools with higher levels of bushfire impact, especially for reading and numeracy[4].

[4]The National Assessment Program Literacy and Numeracy (NAPLAN) is program of standardised tests administered annually in Australian schools in Years 3, 5, 7 and 9.

However, not all students are equally exposed to learning loss, and this is in part influenced by the developmental stage of a student as well as their education stage. For example, a study into the effects of the Christchurch Earthquakes (2010–11) on senior secondary students' attainment of school certificates found no significant difference between disaster-affected and non-disaster-affected students two years after the event (Beaglehole et al., 2017). However, Gibbs et al. (2019) caution that very little research investigates the longer term consequences of crises and disasters on learning. Indeed, the Gibbs et al. study, which looked at impacts up to four years after a bushfire event, found depressed achievement results, which corresponded with the different levels of bushfire affectedness. Gibbs et al. (2019) argue "previous findings in the literature of no impact within a 3-year period of the disaster event should be reviewed" (p. 1410). Another study, which took a long-term view of the impacts of a disaster, involved a 20-year follow-up of children whose education was interrupted by a bushfire. The findings demonstrated that bushfire affected students were less likely to extend their education and careers than the comparison student group (McFarlane & Van Hooff, 2009).

The difference in impact is also, in part, due to the students' home context as well as their socio-economic status. Picou and Marshall (2007) found the resettlement of some Hurricane Katrina affected students to unstable living conditions and unreliable access to transportation had a negative impact on academic attainment. The influence of the home environment on academic achievement following crisis is also illustrated by the educational outcomes following Hurricane Katrina in 2005. Relocation to new schools aided some students' academic development, but inequitable access to education was a key issue for students (Picou & Marshall, 2007). A study by Lamb et al. (2013) into academic achievement in mathematics for students affected by Hurricane Katrina found negative changes in test scores for some sub-groups of students (although not in all cases), with the study's authors concluding that the hurricane "had the most profound effect on mathematics achievement in Mississippi's poor/rural schools" (Lamb et al., 2013, p. 83). A study of the longer-term impacts of the disaster on academic performance on one cohort of relocated students found that while they initially experienced declines in their test scores, by the third and fourth years following the hurricane, they recorded improvements in scores (Sacerdote, 2012).

The COVID-19 pandemic is a cautionary tale of the ongoing and nefarious impacts of school closures on student achievement and wellbeing. The COVID-19 pandemic school closures highlight how, in the absence of on-site schooling, learning is brokered by the home environment. This makes it difficult for schools to redress inequities. Broader systematic issues such as socio-economic disparities amongst households and patterns of advantage and disadvantage (Bourdieu, 1984) influence how students learn at home during school closures. Students who were already experiencing severe social and educational exclusion or faced persistent disadvantage were at greater risk of learning loss while learning from home (Brown et al., 2020; Clinton, 2020; McDaid et al., 2020). This is echoed by Bayrakdar and Guveli (2020) who found that students who are already having trouble obtaining key resources for engaging with school, such as food and uniforms, were at greater risk of learning loss. The capacity of students to "find their way around closed doors" (OECD, 2021, p. 3) was thus conditioned by their socioeconomic and educational context. However, crises also give rise to opportunities for innovations in learning. Investigating COVID-19 school closures in Norway, Bubb and Jones (2020) identified that vulnerable students academically performed better at home due to more follow up from teachers, and partnerships between teachers and parents/carers.

Unsurprisingly, scholars worldwide are concerned that COVID-19 related school closures have worsened pre-existing educational inequalities (Bessell, 2021; Brown et al., 2020; Cullinane & Montacute, 2020; Drane et al., 2020; Lancker & Parolin, 2020; OECD, 2021; United Nations, 2020). This has significant global ramifications. Educational attainment is one of the strongest predictors of a nation's health and wealth (Viner et al., 2020). Educational inequities exposed by coronavirus are expected to persist in the mid- to long-term with consequences for individuals and nations (Bayrakdar & Guveli, 2020; Clinton, 2020; Sonnemann & Goss, 2020). Using data from 157 countries, Azevedo et al. (2020) simulated impacts of learning loss resulting from the COVID-19 pandemic school closures on school-aged children, demonstrating a 0.6-year reduction of schooling would reduce a child's schooling life from 7.9 to 7.3 years and "each student from today's cohort in primary and secondary school could face, on average, a reduction of $872 in yearly earnings. This is approximately equivalent to $16,000 over a student's work life at present value" (p. 4).

Therefore, school closures have the potential to disrupt students' learning and their academic progress in the short and long term (Davis et al., 2021; Gibbs et al., 2019). Delayed recovery can have long-term impacts on educational attainment, and recovery efforts must be generous with time to allow students to regain their connection with school and education

Section 3: Lessons Learned from Crises

In this section, we focus on broader disaster management response processes and systems that also influence access to education. We argue that while a focus on basic needs and crisis containment is understandable, it risks overlooking the broader educational, social, and emotional needs of children and young people. We also examine how the role of the school expands during crises, which puts pressure on educators. Together, these issues exacerbate the risk of learning loss and poorer mental health and wellbeing. Lastly, we highlight how this is, in part, linked to the exclusion of children and young people from disaster responses.

Policy Responses Need to Go Beyond a Focus on Basic Needs and Crisis Containment

The implementation of disaster response policies like school closures and school relocation tends to focus on immediate basic needs, such as safety and the provision of shelter and food. This is understandable in the short term (see Maslow, 1954; Noltemeyer et al., 2021). However, some research emphasises the role of school closures for health and safety, rather than in relation to how school closures might impact on learning, mental health and wellbeing, and life opportunities (Howard-Jones et al., 2022). Notably, studies relating to schooling and the influenza pandemic of 1918–19 and the H1N1 "swine flu" pandemic have largely focused on epidemiological impacts, for example considering the extent to which school closures mitigate transmission (House et al., 2011; Stern et al., 2009). Copeland et al. (2013) found closing schools significantly reduces transmission during the peak of an influenza outbreak, and research by Garbe et al. (2020) and Kneale et al. (2020) shows that school closures are a useful non-pharmaceutical intervention for management of viral epidemics. Some literature has considered the social and economic impacts of the H1N1 "swine flu" pandemic school closure on children's social interactions (see Chen et al., 2011), but the educational impacts of influenza-induced school closures on educational attainment have been under-researched. In contrast, there is much more research in relation to how the COVID-19 pandemic school closures have impacted on students (beyond their physical health), perhaps due to the extensive (world-wide) and extended (long duration) nature of both the pandemic itself and the school lockdowns.

School closure as disaster response policy is typically enacted in response to viral pandemics, like COVID-19 or influenza. This is because the recommendation to close schools is based on the epidemiological benefits of school closures to mitigate transmission of the virus (House et al., 2011; Stern et al., 2009). Emerging data suggests newer variants of the virus pose a greater threat to children and young people, highlighting both the potential for additional and ongoing school closures as well as the need to centre educational equity in crisis responses. However, Joseph and Fahey (2020) have argued school closures are "bad policy" in that they inadequately balance health concerns with economic, social, and educational priorities. Moreover, being treated as a potential virus-spreader rather than a human being is deeply distressing, as captured by this comment from a child in an international survey: "Stop telling people we are vectors it hurts our feelings and make us sad I want to see my friends and my nana and grandad" (Boy, 8, Ireland, in Terre des Hommes, 2020, p. 13).

School closures are not enacted in isolation but are normally part of a society-wide shift to accommodate the demands of crisis and disaster. Parents and carers are also impacted, which exacerbates how crisis and disaster affect children. For example, examining H1N1 swine flu (2009) school closures, Chen et al. (2011) concluded that short-term school closures increased social and economic pressure on households and schools, with unspecified flow-on effects to children. Brown et al. (2011) examined the social and economic cost of H1N1 swine flu (2009) school closures further and concluded:

> Closing schools during the 2009 H1N1 epidemic could have resulted in substantial costs to society as the potential costs of lost productivity and childcare could have far outweighed the cost savings in preventing influenza cases. (p. 1)

The COVID-19 pandemic is further illustrative of the economic pressure school closures place on families. Azevedo et al. (2020, p. 2) note that the pandemic created a "twin shock" for education: the rapid shift to school closures occured alongside the initial recessionary impacts to global economies.

The management of COVID-19 school closures around the world also shows the uncertainty that is caused by a lack of policy coordination. Disagreements within and between nations on how to best manage COVID-19 infection rates led to confusion and ad hoc decision making.

Within Australia, these differences in education system responses potentially led to inequity, disparity and inconsistency in schooling, curriculum delivery and the experiences of students, families, and teachers. In an initial constructive move, in Australia the federal Minister for Education commissioned university researchers to produce rapid research reports across a range of education sectors in April/May 2020[5]. This was a proactive example of mobilizing resources to inform policy development in responding to a crisis—however, ultimately it did not resolve political differences in decision-making between state jurisdictions, which hold the main responsibility for education in Australia. In reporting findings from previous studies of education system responses to viral outbreaks, Cauchemez et al. (2014) cautioned that different school systems coexist, which impacts on the capacity for school systems to respond to crisis in a locally informed way.

Schools Need to be Enabled to Take on Crisis Response Management

While a schools' primary role is the education (broadly conceived) of students, schools and school systems can also play a central role in crisis and disaster response management. Schools are a vital, often trusted, point of connection for communities and are expected by many to lead the community to recovery (Cahill et al., 2020a). During such events, schools' roles expand beyond their usual scope, often becoming crisis management sites with little or no notice. Schools are expected to serve as a key pillar in the recovery effort for students and the broader community (Davis et al., 2021; Lai et al., 2016; Mutch, 2020).

Experiences following the 2011 Christchurch Earthquake are illustrative of the central role that schools serve in their communities during an emergency (Mutch, 2015; Mutch et al., 2014). Here, schools' roles expanded beyond their usual scope of learning to encompass disaster response and recovery support, and school leaders became "crisis managers" (Mutch, 2015). These roles can impose new expectations and responsibilities on school leadership and school staff, for which they are unlikely to be specifically prepared through pre-service and in-service learning. This has implications for future professional development of school leaders. Additional expectations can have particularly negative consequences

[5] See: https://www.dese.gov.au/about-us/announcements/new-research-examine-potential-impact-remote-learning.

in schools in disadvantaged communities which may face more pressure to take on these extra roles (since these communities may have fewer other resources) while at the same time these schools often are already under-resourced themselves.

The work of schools in disaster and crisis management extends long after the disruptive event. This includes influencing policy change, assisting with community rebuilding, and expanding the services offered by schools to support student wellbeing. For instance, policy outcomes in response to the Great East Japan Earthquake (2011) highlight how school responses to disaster can influence national disaster management frameworks. Tragically, 74 students and 10 teachers from Okawa Elementary School in Ishinomaki City lost their lives to the tsunami (Sakurai et al., 2020). The tragedy was investigated; many families sued the Prefecture and City governments, and it was found that communications issues caused by power outages meant students remained on the school grounds rather than escaping to higher grounds nearby to avoid floodwaters. The legal ruling stated that the school had neglected its obligations to clarify evacuation procedures in its risk management manual (Sakurai et al., 2020). It further ruled that if the school disaster contingency plan had been updated, the deaths of the school children could have been prevented.

Prior to the 2011 tsunami, educational disaster management guidelines were informed by lessons from earlier disasters like the Great Hanshin-Awaji Earthquake (1995) (Fujioka & Sakakibara, 2018). The 2011 disaster draws attention to the responsibilities borne by governments and schools to provide appropriate disaster planning and advice for responding to crises. Following the event, disaster risk reduction education became a focus for the Japanese government, and participatory disaster risk management that involves the local community, has been identified as a preferable means of effectively communicating risk (Shaw et al., 2021). Schools teach students disaster management drills as part of their curriculum (Fujioka & Sakakibara, 2018).

The role of schools in disaster and crisis management can also extend to supporting community rebuilding. In Iran, a program of "Safe Schools-Resilient Communities" (Amini Hosseini & Izadkhah, 2020) promotes broader public participation in disaster management at community level. This type of approach shifts from a focus on a singular crisis or disaster (such as earthquake preparedness), and instead focuses on increasing community resilience to develop capacities to respond to any crisis or disaster.

The important role that schools can play in assisting with disaster management policy development is exemplified by the recovery from Hurricane Katrina. The policy response to Hurricane Katrina illustrates how schools and local government can work together in crisis and disaster management. In New Orleans, the local government implemented systemic changes to education to build resilience in school communities. After Hurricane Katrina, education in New Orleans was decentralised to remodel the urban school district (Thomas et al., 2013). Findings from education research conducted post-Katrina indicate that school-based interventions and positive school environments where there is an ethos of cooperation can mitigate disaster related impacts on academic performance (Barrett et al., 2012; Sacerdote, 2008).

The gradual reopening of schools in the United States following the influenza pandemic (1918–1919) highlights the importance of community collaboration to getting students back in school. Battenfeld (2020) points to three lessons of that pandemic that would assist with keeping students in school: investing in school nurses as a model of preventative health; developing strong collaborative partnerships between schools and emergency services and linking education with other priorities like health.

Children and Young People Need to be Involved in Crisis Responses

More work needs to be done to empower children's voices in response to disaster and crisis (Amri et al., 2018). Gibbs et al. (2013, p. 132) describe how, following crisis, students are largely positioned as victims: "constrained to play the role of spectator, seen as having little to say of value about the new world in which they live, and how they hope it might be changed." Reflecting on the response to the COVID-19 pandemic, Cuevas-Parra (2020) argues for a re-think in research in order to position children and young people as co-contributors.

Over the last three decades, research has increasingly called for the inclusion of children and young people in disaster management (Fothergill & Peek, 2015; Freeman et al., 2015). Peek (2008) suggests involving children in disaster management and recovery is central to building community resilience. Research into the Christchurch Earthquakes (2010–11) by Freeman et al. (2015) engaged with ninety-four children affected by the disaster. Crucially, Freeman and colleagues repositioned children "not merely as vulnerable recipients in need of protection" but "as competent social actors with voice and agency, capable of understanding and acting in response to disasters" (p. 104).

The Great East Japan Earthquake (2011) disaster facilitated a global policy shift, signposted by the 2015–2030 Sendai Framework for Disaster Risk Reduction (United Nations, 2015). Central to this, is the involvement of children in decision making: "children and youth are agents of change and should be given the space and modalities to contribute to disaster risk reduction, in accordance with legislation, national practice and educational curricula" (United Nations, 2015, p. 23).

The 2019–2020 Australian bushfires led to the establishment of a peak body for young people in Victoria to create resources to ensure organisations "include and engage young people effectively and meaningfully in disaster response" (Youth Affairs Council of Victoria, n.d., n.p.). They include factsheets on how to consult with young people and how to include them on committees; a report on young people's experience of bushfires and recovery; a resource for young people with disabilities; and a Youth Focused Disaster Recovery Protocol.

Also in Australia, children and young people can contact "Kids Helpline", a free, confidential 24/7 online and phone counselling service for young people aged 5 to 25. The helpline received many more calls as the pandemic first started to affect Australia, in early 2020. In recognition of the importance of listening to children and young people for how governments and organisations respond to the pandemic, a report has been published outlining the impact of the COVID-19 pandemic on children and young people based on what these callers to the helpline said (Nicolson et al., 2020). The top five concerns raised were:

1. Mental health concerns resulting from the COVID-19 pandemic
2. Social isolation
3. Education impacts
4. Impacts on family life
5. Changes to plans and usual activities. (p. 12)

Internationally, a large number of organisations, led by Terre des Hommes, in collaboration with the United Nations launched a survey for children and young people to garner their views about the pandemic during 2020. The survey was co-designed with children and young people, and they also contributed to data analysis and interpretation (Terre des Hommes, 2020). In total 26,258 children aged between 8–17 from 137 countries contributed to the survey, with views about a large range of topics. Importantly, "38% don't think their government is listening to children when making decisions about COVID-19" (Terre des Hommes, 2020, p. 7). Quotes in the report from children and young people on this topic include:

> I would tell politicians when they are making laws to do that with the heart of mothers and not of politicians. (Girl, 12, Bolivia, in Terre des Hommes, 2020, p. 7)

> A country is not only made up of adults, we children exist and should be heard as well. (Girl 14, Costa Rica, in Terre des Hommes, 2020, p. 10)

> We should be allowed to voice our opinions on when schools should reopen. It is our future so it should be our choice (Girl, 15, South Africa, in Terre des Hommes, 2020, p. 25)

These examples of research and action all demonstrate how children and young people can actively contribute to the management of, and recovery from, disaster and crisis. The inclusion of children and young people in response to disaster and crisis will be increasingly vital to bolstering community resilience, at a global level. Rees et al. (2021) argue that children have a fundamental right to be involved in the management of crises (see further Mort et al., 2020). Focusing on the ongoing unfolding climate crisis, Rees et al. (2021) posits that involving children in disaster and crisis is in the "best interests of the child" (p. 109). Bessell (2021) advocates for including children's voices in the dual process of recovering and innovating education (also see Cahill et al., 2020b).

Conclusion

Crisis and disaster spur change, opening opportunities for a review of education. Echoing Davis et al. (2021) following disaster "the path for schools to find a new normal is not well documented" (p. 1), however, past, and present crises collectively offer insight into key challenges and opportunities for protecting equitable access to education for children and young people during, and following, times of crisis and disaster.

The COVID-19 pandemic has spurred innovation within the education sector (United Nations, 2020; Zhao & Watterson, 2021). Educators rapidly upskilled for digital learning and there is potential for blended or hybrid teaching and learning provisions to become more common (Adedoyin & Soykan, 2020). Zhao and Watterson (2021) argue the pandemic brought forward necessary innovations which had been dormant in the sector, including personalised curriculum, student-centred inquiry-based learning, and hybrid learning that is both synchronous and asynchronous. In Australia, new funding schemes are connecting tutors with children in need to redress some of the learning lost from the COVID-19 school closures (Sonnemann & Goss, 2020; The Smith Family, 2021). Sahlberg (2020) suggests that the experiences during COVID-19 will increase the willingness among those in power to make education more equitable. The next three chapters focus specifically on what we have learnt about education and equity from our own research in Australia and international literature in the context of the COVID-19 pandemic.

References

Adams, R. (2021, April 4). Teaching union urges schools to stop 'parent bombing' of online classes. *The Guardian*. Retrieved April 4, 2021, from https://www.theguardian.com/education/2021/apr/03/teaching-union-urges-schools-to-stop-parent-bombing-of-online-classes?CMP=Share_iOSApp_Other

Adedoyin, O. B., & Soykan, E. (2020). Covid-19 pandemic and online learning: The challenges and opportunities. *Interactive Learning Environments*, 1–13. https://doi.org/10.1080/10494820.2020.1813180

Ager, P., Eriksson, K., Karger, E., Nencka, P., & Thomasson, M. A. (2020, December). *School closures during the 1918 flu pandemic* (NBER Working Paper No. 28246). National Bureau of Economic Research. Retrieved July 8, 2021, from https://www.nber.org/system/files/working_papers/w28246/w28246.pdf

Amini Hosseini, K., & Izadkhah, Y. O. (2020). From "earthquake and safety" school drills to "safe school-resilient communities": A continuous attempt for promoting community-based disaster risk management in Iran. *International Journal of Disaster Risk Reduction, 45*, 101512. https://doi.org/10.1016/j.ijdrr.2020.101512

Amri, A., Haynes, K., Bird, D. K., & Ronan, K. (2018). Bridging the divide between studies on disaster risk reduction education and child-centred disaster risk reduction: A critical review. *Children's Geographies, 16*(3), 239–251. https://doi.org/10.1080/14733285.2017.1358448

Ardagh, M. W., Richardson, S. K., Robinson, V., Than, M., Gee, P., Henderson, S., Khodaverdi, L., McKie, J., Robertson, G., Schroeder, P. P., & Deely, J. M. (2012). The initial health-system response to the earthquake in Christchurch, New Zealand, in February, 2011. *The Lancet, 379*(9831), 2109–2115. https://doi.org/10.1016/S0140-6736(12)60313-4

Armitage, R., & Nellums, L. B. (2020). Considering inequalities in the school closure response to COVID-19. *The Lancet Global Health, 8*(5), E644. https://doi.org/10.1016/S2214-109X(20)30116-9

Azevedo, J. P., Hasan, A., Goldemberg, D., Iqbal, S. A., & Geven, K. (2020, June 18). Simulating the potential impacts of COVID-19 school closures on schooling and learning outcomes: A set of global estimates. *World Bank Group: Education Global Practice*. Retrieved February 2, 2021, from https://www.worldbank.org/en/topic/education/publication/simulating-potential-impacts-of-covid-19-school-closures-learning-outcomes-a-set-of-global-estimates

Barrett, E. J., Ausbrooks, C. Y. B., & Martinez-Cosio, M. (2012). The tempering effect of schools on students experiencing a life-changing event: Teenagers and the Hurricane Katrina evacuation. *Urban Education, 47*(1), 7–31. https://doi.org/10.1177/0042085911416011

Battenfeld, M. (2020, June 16). 3 lessons from how schools responded to the 1918 pandemic worth heeding today. *The Conversation*. Retrieved January 15, 2021, from https://theconversation.com/3-lessons-from-how-schools-responded-to-the-1918-pandemic-worth-heeding-today-138403

Bayrakdar, S., & Guveli, A. (2020). *Inequalities in home learning and schools' provision of distance teaching during school closure of COVID-19 lockdown in the UK* (ISER Working Paper Series No. 2020-09). Institute for Social and Economic Research. Retrieved January 18, 2021, from https://www.econstor.eu/bitstream/10419/227790/1/1703719352.pdf

Beaglehole, B., Bell, C., Frampton, C., & Moor, S. (2017). The impact of the Canterbury earthquakes on successful school leaving for adolescents. *Australian and New Zealand Journal of Public Health, 41*(1), 70–73. https://doi.org/10.1111/1753-6405.12625

Beasy, K., Murphy, C., Hicks, D., & Muir, T. (2020). *Experiences of parents and caregivers supporting children while learning from home* (Share Your Experiences of COVID-19, Report No. 9). University of Tasmania. Retrieved February 4, 2021, from https://blogs.utas.edu.au/isc/files/2020/05/Experiences-of-parents-and-caregivers-supporting-children-while-learning-from-home.-Report-9.pdf

Bessell, S. (2021). The impacts of COVID-19 on children in Australia: Deepening poverty and inequality. *Children's Geographies, 20*(4), 448–458. https://doi.org/10.1080/14733285.2021.1902943

Bourdieu, P. (1984). *Distinction: A social critique of the judgement of taste*. Harvard University Press.

Bourke, R., O'Neill, J., McDowall, S., Dacre, M., Mincher, N., Narayanan, V., Overbye, S., & Tuifagalele, R. (2021). *Children's informal learning at home during COVID-19 lockdown*. New Zealand Council for Educational Research and Massey University. Retrieved October 1, 2021, from https://www.nzcer.

org.nz/system/files/NZC_Children%27s%20informal%20learning%20Final%2023-8-21.pdf

Bronfenbrenner, U. (1974). Developmental research, public policy, and the ecology of childhood. *Child Development, 45*(1), 1–5. https://doi.org/10.2307/1127743

Bronfenbrenner, U. (1977). Toward an experimental ecology of human development. *The American Psychologist, 32*(7), 513–531. https://doi.org/10.1037/0003-066X.32.7.513

Bronfenbrenner, U. (1995). The bioecological model from a life course perspective: Reflections of a participant observer. In P. Moen, K. Lüscher, & G. H. Elder Jr. (Eds.), *Examining lives in context: Perspectives on the ecology of human development* (pp. 599–618). American Psychological Association.

Brown, N., Riele, K. T., Shelley, B., & Woodroffe, J. (2020). *Learning at home during COVID-19: Effects on vulnerable young Australians* (Independent Rapid Response Report). Peter Underwood Centre for Educational Attainment, University of Tasmania. Retrieved May 26, 2020, from https://www.utas.edu.au/__data/assets/pdf_file/0008/1324268/Learning-at-home-during-COVID-19-updated.pdf

Brown, S. T., Tai, J. H., Bailey, R. R., Cooley, P. C., Wheaton, W. D., Potter, M. A., Voorhees, R. E., LeJeune, M., Grefenstette, J. J., Burke, D. S., McGlone, S. M., & Lee, B. Y. (2011). Would school closure for the 2009 H1N1 influenza epidemic have been worth the cost?: A computational simulation of Pennsylvania. *BMC Public Health, 11*, 353. https://doi.org/10.1186/1471-2458-11-353

Bryant, R. A., Gibbs, L., Gallagher, H. C., Pattison, P., Lusher, D., MacDougall, C., Harms, L., Block, K., Sinnott, V., Ireton, G., Richardson, J., & Forbes, D. (2018). Longitudinal study of changing psychological outcomes following the Victorian Black Saturday bushfires. *Australian & New Zealand Journal of Psychiatry, 52*(6), 542–551. https://doi.org/10.1177/0004867417714337

Bubb, S., & Jones, M.-A. (2020). Learning from the COVID-19 home-schooling experience: Listening to pupils, parents/carers and teachers. *Improving Schools, 23*(3), 209–222. https://doi.org/10.1177/1365480220958797

Cahill, H., Dadvand, B., Shlezinger, K., Romei, K., & Farrelly, A. (2020a). *Natural disasters and pandemics: Supporting student and teacher wellbeing post emergency* (MGSE Industry Report No. 3). Youth Research Centre, Melbourne Graduate School of Education, the University of Melbourne. Retrieved August 12, 2020, from https://education.unimelb.edu.au/__data/assets/pdf_file/0006/3453036/UoM-MGSE-Industry-Report-Post-Emergency_Web_FA.pdf

Cahill, H., Shlezinger, K., Romei, K., & Dadvand, B. (2020b). *Research-informed approaches to supporting student wellbeing post-disaster*. Youth Research Centre.

Caley, P., Philp, D. J., & McCracken, K. (2008). Quantifying social distancing arising from pandemic influenza. *Journal of the Royal Society Interface, 5*(23), 631–639. https://doi.org/10.1098/rsif.2007.1197

Cauchemez, S., Van Kerkhove, M. D., Archer, B. N., Cetron, M., Cowling, B. J., Grove, P., Hunt, D., Kojouharova, M., Kon, P., Ungchusak, K., Oshitani, H., Pugliese, A., Rizzo, C., Saour, G., Sunagawa, T., Uzicanin, A., Wachtel, C., Weisfuse, I., Yu, H., & Nicoll, A. (2014). School closures during the 2009 influenza pandemic: National and local experiences. *BMC Infectious Diseases, 14*, 207. https://doi.org/10.1186/1471-2334-14-207

Chen, W. C., Huang, A. S., Chuang, J. H., Chiu, C. C., & Kuo, H. S. (2011). Social and economic impact of school closure resulting from pandemic influenza A/H1N1. *Journal of Infection, 62*(3), 200–203. https://doi.org/10.1016/j.jinf.2011.01.007

Chowell, G., Ammon, C. E., Hengartner, N. W., & Hymann, J. M. (2006). Transmission dynamics of the great influenza pandemic of 1918 in Geneva, Switzerland: Assessing the effects of hypothetical interventions. *Journal of Theoretical Biology, 241*(2), 193–204. https://doi.org/10.1016/j.jtbi.2005.11.026

Chrysanthos, N. (2020, January 17). NSW to spend $20 million on bushfire recovery for 178 schools. *The Sydney Morning Herald*. Retrieved June 28, 2021, from https://www.smh.com.au/national/nsw/nsw-to-spend-20-million-on-bushfire-recovery-for-178-schools-20200116-p53s04.html

Clifton, C., Bruneau, M., MacRae, G., Leon, R., & Fussell, A. (2011). Steel structures damage from the Christchurch earthquake series of 2010 and 2011. *Bulletin of the New Zealand Society for Earthquake Engineering, 44*(4), 297–318. https://doi.org/10.5459/bnzsee.44.4.297-318

Clinton, J. (2020). *Supporting vulnerable children in the face of a pandemic: A paper prepared for the Australian Government Department of Education, Skills and Employment*. Centre for Program Evaluation, Melbourne Graduate School of Education, University of Melbourne.

Copeland, D. L., Basurto-Davila, R., Chung, W., Kurian, A., Fishbein, D. B., Szymanowski, P., Zipprich, J., Lipman, H., Cetron, M. S., Meltzer, M. I., & Averhoff, F. (2013). Effectiveness of a school district closure for pandemic influenza A (H1N1) on acute respiratory illnesses in the community: A natural experiment. *Clinical Infection Diseases, 56*(4), 509–516. https://doi.org/10.1093/cid/cis890

Cuevas-Parra, P. (2020). Co-Researching with children in the time of COVID-19: Shifting the narrative on methodologies to generate knowledge. *International Journal of Qualitative Methods, 19*, 1–12. https://doi.org/10.1177/1609406920982135

Cullinane, C., & Montacute, R. (2020). *COVID-19 and Social Mobility Impact Brief #1: School Shutdown*. The Sutton Trust. Retrieved August 15, 2020, from https://www.suttontrust.com/wp-content/uploads/2021/01/School-Shutdown-Covid-19.pdf

Davis, C. R., Cannon, S. R., & Fuller, S. C. (2021). The storm after the storm: the long-term lingering impacts of hurricanes on schools. *Disaster Prevention and Management: An International Journal, 30*(3), 264–278. https://doi.org/10.1108/DPM-03-2020-0055

Dickinson, H., & Yates, S. (2020). *More than isolated: The experience of children and young people with disability and their families during the COVID-19 pandemic.* Children and Young People with Disability Australia. Retrieved November 19, 2020, from https://www.cyda.org.au/images/pdf/covid_report_compressed_1.pdf

Drane, C., Vernon, L., & O'Shea, S. (2020). *The impact of 'learning at home' on the educational outcomes of vulnerable children in Australia during the COVID-19 pandemic.* National Centre for Student Equity in Higher Education for the Department of Education, Skills and Employment, Curtin University. Retrieved May 26, 2020, from https://www.ncsehe.edu.au/wp-content/uploads/2020/04/NCSEHE_V2_Final_literaturereview-learningathome-covid19-final_30042020.pdf

Fothergill, A., & Peek, L. (2006). Surviving catastrophe: A study of children in Hurricane Katrina. *Learning From Catastrophe: Quick Response Research in the Wake of Hurricane Katrina, 40*, 97–129.

Fothergill, A., & Peek, L. (2015). *Children of Katrina.* University of Texas Press.

Freeman, C., Nairn, K., & Gollop, M. (2015). Disaster impact and recovery: What children and young people can tell us. *Kōtuitui: New Zealand Journal of Social Sciences Online, 10*(2), 103–115. https://doi.org/10.1080/1177083X.2015.1066400

Fujioka, T., & Sakakibara, Y. (2018). School education for disaster risk reduction in Japan after the 2011 Great East Japan Earthquake and Tsunami (GEJET). *Terrae Didatica, 14*(3), 313–319. https://doi.org/10.20396/td.v14i3.8653531

Garbe, A., Ogurlu, U., Logan, N., & Cook, P. (2020). COVID-19 and remote learning: Experiences of parents with children during the pandemic. *American Journal of Qualitative Research, 4*(3), 45–65.

Gibbs, L., Mutch, C., O'Connor, P., & MacDougall, C. (2013). Research with, by, for and about children: Lessons from disaster contexts. *Global Studies of Childhood, 3*(2), 129–141.

Gibbs, L., Nursey, J., Cook, J., Ireton, G., Alkemade, N., Roberts, M., Gallagher, H. C., Bryant, R., Block, K., Molyneaux, R., & Forbes, D. (2019). Delayed disaster impacts on academic performance of primary school children. *Child Development, 90*(4), 1402–1412. https://doi.org/10.1111/cdev.13200

Goto, A., Lloyd Williams, A., Okabe, S., Koyama, Y., Koriyama, C., Murakami, M., Yui, Y., & Nollet, K. E. (2022). Empowering children as agents of change to Foster Resilience in Community: Implementing "Creative Health" in primary schools after the Fukushima Nuclear Disaster. *International Journal of Environmental Research and Public Health, 19*(6), 3417. https://doi.org/10.3390/ijerph19063417

Graham, A., Powell, M. A., Thomas, N., & Anderson, D. (2017). Reframing 'well-being' in schools: The potential of recognition. *Cambridge Journal of Education, 47*(4), 439–455. https://doi.org/10.1080/0305764X.2016.1192104

Grewenig, E., Lergetporer, P., Werner, K., Woessmann, L., & Zierow, L. (2020). *COVID-19 and educational inequality: How school closures affect low-and high-achieving students* (Discussion Paper No. 260). Ludwig-Maximilians-Universität München und Humboldt-Universität zu Berlin, Collaborative Research Center Transregio 190. Retrieved September 27, 2021, from https://www.econstor.eu/bitstream/10419/233484/1/1755164874.pdf

Hall, V., Cathro, G., Winter, M., & Winter, J. (2012). *Evaluative study of co-located schools established following the Christchurch earthquake*. Ministry of Education, New Zealand. Retrieved June 28, 2021, from http://shapingeducation.govt.nz/wp-content/uploads/2012/12/989-Co-located-schools-report-Final.pdf

Harms, L., Gibbs, L., Ireton, G., MacDougall, C., Brady, K., Kosta, L., Block, K., Baker, E., Gallagher, H. C., Kellett, C., Forbes, D., & Bryant, R. (2021). Stressors and supports in postdisaster recovery: Experiences after the Black Saturday bushfires. *Australian Social Work, 74*(3), 332–347.

Haug, N., Geyrhofer, L., Londel, A., Dervic, E., Desvars-Larrive, A., Loreto, V., Pinior, B., Thurner, S., & Klimek, P. (2020). Ranking the effectiveness of worldwide COVID-19 government interventions. *Nature Human Behaviour, 4*, 1303–1312. https://doi.org/10.1038/s41562-020-01009-0

House, T., Baguelin, M., Van Hoek, A. J., White, P. J., Sadique, Z., Eames, K., Read, J. M., Hens, N., Melegaro, A., Edmunds, W. J., & Keeling, M. J. (2011). Modelling the impact of local reactive school closures on critical care provision during an influenza pandemic. *Proceedings of the Royal Society B: Biological Sciences, 278*(1719), 2753–2760. https://doi.org/10.1098/rspb.2010.2688

Howard-Jones, A. R., Bowen, A. C., Danchin, M., Koirala, A., Sharma, K., Yeoh, D. K., Burgner, D. P., Crawford, N. W., Goeman, E., Gray, P. E., Hsu, P., Kuek, S., McMullan, B. J., Tosif, S., Wurzel, D., & Britton, P. N. (2022). COVID-19 in children: I. Epidemiology, prevention and indirect impacts. *Journal of Paediatrics and Child Health, 58*(1), 39–45. https://doi.org/10.1111/jpc.15791

Johnson, V. A., & Ronan, K. R. (2014). Classroom responses of New Zealand school teachers following the 2011 Christchurch earthquake. *Natural Hazards, 72*(2), 1075–1092. https://doi.org/10.1007/s11069-014-1053-3

Joseph, B., & Fahey, G. (2020). *Pain without gain: Why school closures are bad policy* (Policy Paper No. 28). The Centre for Independent Studies. Retrieved November 19, 2020, from https://www.cis.org.au/app/uploads/2020/05/pp28.pdf?

Klaiman, T., Kraemer, J. D., & Stoto, M. A. (2011). Variability in school closure decisions in response to 2009 H1N1: A qualitative systems improvement analysis. *BMC Public Health, 11*(1), 73. https://doi.org/10.1186/1471-2458-11-73

Kneale, D., O'Mara-Eves, A., Rees, R., & Thomas, J. (2020). School closure in response to epidemic outbreaks: Systems-based logic model of downstream impacts. *F1000 Research, 9*(1), 352. https://doi.org/10.12688/f1000research.23631.1

Kousky, C. (2016). Impacts of natural disasters on children. *The Future of Children, 26*(1), 73–92.

Lai, B. S., Esnard, A.-M., Lowe, S. R., & Peek, L. (2016). Schools and disasters: Safety and mental health assessment and interventions for children. *Child and Family Disaster Psychiatry, 18*(109), 109–118. https://doi.org/10.1007/s11920-016-0743-9

Lamb, J., Gross, S., & Lewis, M. (2013). The Hurricane Katrina effect on mathematics achievement in Mississippi. *School Science and Mathematics, 113*(2), 80–93. https://doi.org/10.1111/ssm.12003

Lamb, S. (2020). *Impact of learning from home on educational outcomes for disadvantaged children.* Centre for the International Research on Education Systems, Victoria University. Retrieved May 26, 2020, from https://apo.org.au/node/303564

Lancker, W. V., & Parolin, Z. (2020). COVID-19, school closures, and child poverty: A social crisis in the making. *The Lancet Public Health, 5*(5), E243–E244. https://doi.org/10.1016/S2468-2667(20)30084-0

Lee, J. (2020). Mental health effects of school closures during COVID-19. *The Lancet Child & Adolescent Health, 4*(6), P421. https://doi.org/10.1016/S2352-4642(20)30109-7

Lieber, M. (2017). Assessing the mental health impact of the 2011 Great Japan Earthquake, Tsunami, and radiation disaster on elementary and middle school children in the Fukushima Prefecture of Japan. *PLOS One, 12*(1), 1–14. https://doi.org/10.1371/journal.pone.0170402

Maslow, A. (1954). *Motivation and personality.* Harper & Row.

Mathieu, E., Ritchie, H., Ortiz-Ospina, E., Roser, M., Hasell, J., Appel, C., Giattino, C., & Rodés-Guirao, L. (2021). A global database of COVID-19 vaccinations. *Nature Human Behaviour, 5*(7), 947–953. https://doi.org/10.1038/s41562-021-01122-8

McDaid, L., Tomaszewski, W., Staton, S., Povey, J., Thorpe, K., Vaithianathan, R., Ridgway, T., Diffey, B., & Western, M. (2020). *Who are the students most at risk of falling being in their learning?* (Pillar 1 Report). Institute for Social Science Research, University of Queensland. Retrieved June 16, 2021, from https://issr.uq.edu.au/files/17185/Ramsay%20Pillar%201_Full%20Report.pdf

McFarlane, A. C., & Van Hooff, M. (2009). Impact of childhood exposure to a natural disaster on adult mental health: 20-year longitudinal follow-up study.

British Journal of Psychiatry, 195(2), 142–148. https://doi.org/10.1192/bjp.bp.108.054270

McMahon, T. J., Ward, N. L., Pruett, M. K., Davidson, L., & Griffith, E. E. H. (2000). Building full-service schools: Lessons learned in the development of inter-agency collaboratives. *Journal of Educational and Psychological Consultation, 11*(1), 65–92. https://doi.org/10.1207/S1532768XJEPC1101_5

Mooney, M., Tarrant, R., Paton, D., Johnston, D., & Johal, S. (2021). The school community contributes to how children cope effectively with a disaster. *Pastoral Care in Education, 39*(1), 24–47. https://doi.org/10.1080/02643944.2020.1774632

Mort, M., Rodríguez-Giralt, I., & Delicado, A. (Eds.). (2020). *Children and young people's participation in disaster risk reduction: Agency and resilience.* Policy Press.

Moss, G., Webster, R., Harmey, S., & Bradbur, A. (2021). *Unsung heroes: The role of teaching assistants and classroom assistants in keeping schools functioning during lockdown.* International Literacy Centre, UCL Institute of Education. Retrieved April 14, 2021, from http://maximisingtas.co.uk/assets/unsung-heroesfinal.pdf

Mutch, C. (2015). Leadership in times of crisis: Dispositional, relational and contextual factors influencing school principals' actions. *International Journal of Disaster Risk Reduction, 14*(2), 186–194. https://doi.org/10.1016/j.ijdrr.2015.06.005

Mutch, C. (2020). How might research on schools' responses to earlier crises help us in the COVID-19 recovery process? *Set: Research Information for Teachers.* https://doi.org/10.18296/set.0169

Mutch, C., Gibbs, L., O'Connor, P., & MacDougall, C. (2014). Editorial. *Pastoral Care in Education, 32*(1), 1–4. https://doi.org/10.1080/02643944.2014.881599

Nicolson, S., Newell, S., & Collyer, B. (2020). *Impacts of COVID-19 on children and young people who contact Kids Helpline (2020).* Australian Human Rights Commission. Retrieved November 16, 2020, from https://humanrights.gov.au/our-work/childrens-rights/publications/impacts-covid-19-children-and-young-people-who-contact-kids

Noltemeyer, A., James, A. G., Bush, K., Bergen, D., Barrios, V., & Patton, J. (2021). The relationship between deficiency needs and growth needs: The continuing investigation of Maslow's theory. *Child & Youth Services, 42*(1), 24–42. https://doi.org/10.1080/0145935X.2020.1818558

Organisation for Economic Co-Operation and Development. (2021). *The State of School Education One Year into the COVID-19 pandemic.* Retrieved April 4, 2021, from https://www.oecd-ilibrary.org/content/publication/201dde84-en

Peek, L. (2008). Children and disasters: Understanding vulnerability, developing capacities, and promoting resilience — An introduction. *Children, Youth and Environments, 18*(1), 1–29.

Peek, L., Abramson, D. M., Cox, R. S., Fothergill, A., & Tobin, J. (2017). Children and disasters. In H. Rodriguez, W. Donner, & J. Trainor (Eds.), *Handbook of disaster research* (pp. 243–262). Springer. https://doi.org/10.1007/978-3-319-63254-4_13

Peek, L., & Richardson, K. (2010). In their own words: Displaced children's educational recovery needs after Hurricane Katrina [Special issue]. *Disaster Medicine and Public Health Preparedness, 4*(1), S63–S70. https://doi.org/10.1001/dmp.2010.10060910

Picou, J. S., & Marshall, B. K. (2007). Social impacts of Hurricane Katrina on displaced K–12 students and educational institutions in coastal Alabama counties: Some preliminary observations. *Sociological Spectrum, 27*(6), 767–780. https://doi.org/10.1080/02732170701534267

Plyer, A. (2016, August 26). Facts for features: Katrina impact. *The Data Center*. Retrieved June 28, 2021, from https://www.datacenterresearch.org/data-resources/katrina/facts-for-impact/

Potter, S. H., Becker, J. S., Johnston, D. M., & Rossiter, K. P. (2015). An overview of the impacts of the 2010-2011 Canterbury earthquakes. *International Journal of Disaster Risk Reduction, 14*(1), 6–14. https://doi.org/10.1016/j.ijdrr.2015.01.014

Ranghieri, F., & Ishiwatari, M. (2014). *Learning from megadisasters: Lessons from the Great East Japan Earthquake*. Open Knowledge Repository, World Bank Group. Retrieved August 11, 2021, from https://openknowledge.worldbank.org/handle/10986/18864

Rees, N., Barkhof, M., Burdziej, J., Lee, S., & Riley, H. (2021). *The climate crisis is a child rights crisis: Introducing the Children's Climate Risk Index*. United Nations Children's Fund.

Rich, G. (2020, May 13). During the 1918 flu pandemic, at-home learning meant little schoolwork. *The Washington Post*. Retrieved June 19, 2021, from https://www.washingtonpost.com/lifestyle/kidspost/school-during-the-1918-spanish-flu-pandemic/2020/05/12/7d606c90-7f32-11ea-9040-68981f488eed_story.html

Russell, J., Rajanaidu, S., Anderson, P., Best, E., de Lore, D., McKree Jansen, R., Leversha, A., Percival, T., Sinclair, O., & Webb, R. (2022). *Schools and COVID-19 in Aotearoa New Zealand: Keeping schools open as safely as possible.* Retrieved June 2, 2022 from https://cdn.auckland.ac.nz/assets/auckland/news-and-opinion/2022/01/Russell%20et%20al%20%202022%20Schools%20and%20COVID-19%20Keeping%20schools%20open%20(002).pdf

Sacerdote, B. (2008). *When the saints come marching in: Effects of hurricanes Katrina and Rita on student evacuees* (Working Paper No. 14385). National Bureau of Economic Research. https://doi.org/10.3386/w14385

Sacerdote, B. (2012). When the saints go marching out: Long-term outcomes for student evacuees from Hurricanes Katrina and Rita. *American Economic Journal: Applied Economics, 4*(1), 109–135. https://doi.org/10.1257/app.4.1.109

Sahlberg, P. (2020). Does the pandemic help us make education more equitable? *Educational Research for Policy and Practice, 20*, 11–18. https://doi.org/10.1007/s10671-020-09284-4

Sakurai, A., Sato, T., & Murayama, Y. (2020). Impact evaluation of a school-based disaster education program in a city affected by the 2011 great East Japan earthquake and tsunami disaster [Special issue]. *International Journal of Disaster Risk Reduction, 47*, 101632. https://doi.org/10.1016/j.ijdrr.2020.101632

Sandstrom, H., & Huerta, S. (2013). *The negative effects of instability on child development: A research synthesis* (Low Income Working Families, Discussion Paper No. 3). Urban Institute. Retrieved June 19, 2021, from https://www.urban.org/sites/default/files/publication/32706/412899-The-Negative-Effects-of-Instability-on-Child-Development-A-Research-Synthesis.PDF

Shakespeare-Finch, J., Bowen-Salter, H., Cashin, M., Badawi, A., Wells, R., Rosenbaum, S., & Steel, Z. (2020). COVID-19: An Australian perspective [Special issue]. *Journal of Loss and Trauma, 25*(8), 662–672. https://doi.org/10.1080/15325024.2020.1780748

Shaw, R., Sakurai, A., & Oikawa, Y. (2021). New realization of disaster risk reduction education in the context of a global pandemic: Lessons from Japan. *International Journal of Disaster Risk Science, 12*, 568–580. https://doi.org/10.1007/s13753-021-00337-7

Smith, L., & Riley, D. (2012). School leadership in times of crisis. *School Leadership & Management, 32*(1), 57–71. https://doi.org/10.1080/13632434.2011.614941

Sonnemann, J., & Goss, P. (2020). *COVID catch-up: Helping disadvantaged students close the equity gap* (Grattan Institute Report No. 2020-08, June 2020). Grattan Institute, Victoria. Retrieved April 14, 2021, from https://grattan.edu.au/wp-content/uploads/2020/06/COVID-Catch-up-Grattan-School-Education-Report.pdf

Stern, A. M., Cetron, M. S., & Markel, H. (2009). Closing the schools: Lessons from the 1918-19 U.S. influenza pandemic. *Health Affairs, 28*(6), 1066–1078. https://doi.org/10.1377/hlthaff.28.6.w1066

Teague, B., Pascoe, S., & McLeod, R. (2010). *Final Report Summary*. 2009 Victorian Bushfires Royal Commission, Parliament of Victoria. Retrieved June 28, 2021, from http://royalcommission.vic.gov.au/finaldocuments/summary/PF/VBRC_Summary_PF.pdf

Terre des Hommes. (2020). *#CovidUnder19*. https://www.tdh.ch/en/projects/covidunder19

The Smith Family. (2021). *Improving young Australians' literacy and numeracy: The Catch-Up Learning program*. https://www.thesmithfamily.com.au/research/reports/catch-up-learning-pilot

Thomas, D. S., Philips, B. D., Lovekamp, W. E., & Fothergill, A. (Eds.). (2013). *Social vulnerability to disasters*. CRC Press. Retrieved June 28, 2021, from https://www.routledge.com/Social-Vulnerability-to-Disasters/Thomas-Phillips-Lovekamp-Fothergill/p/book/9781466516373

United Nations. (2015). *Sendai Framework for Disaster Risk Reduction 2015–2030. Geneva, Switzerland*

United Nations. (2020). *Policy brief: Education during COVID-19 and beyond*. Retrieved November 16, 2020, from https://www.un.org/development/desa/dspd/wp-content/uploads/sites/22/2020/08/sg_policy_brief_covid-19_and_education_august_2020.pdf

United Nations Educational, Scientific and Cultural Organization. (2021). *Global monitoring of school closures*. Retrieved September 3, 2021, from https://en.unesco.org/covid19/educationresponse#schoolclosures

Viner, R. M., Russell, S. J., Croker, H., Packer, J., Ward, J., Stansfield, C., Mytton, O., Bonell, C., & Booy, R. (2020). School closure and management practices during coronavirus outbreaks including COVID-19: A rapid systematic review. *Lancet Child Adolescent Health, 4*(5), 397–304. https://doi.org/10.1016/S2352-4642(20)30095-X

Wang, G., Zhang, Y., Zhao, J., Zhang, J., & Jiang, F. (2020). Mitigate the effects of home confinement on children during the COVID-19 outbreak. *Lancet, 395*, 945–947.

Yoshikawa, H., Wuermli, A., Britto, P. R., Dreyer, B., Leckman, J. F., Lye, S. J., Ponguta, L. A., Richter, L. M., & Stein, A. (2020). Effects of the global coronavirus disease-2019 pandemic on early childhood development: Short- and long-term risks and mitigating program and policy actions. *The Journal of Pediatrics, 223*, 188–193. https://doi.org/10.1016/j.jpeds.2020.05.020

Youth Affairs Council of Victoria. (n.d.). *Young People and Disasters Resource Hub*. https://www.yacvic.org.au/rural/bushfire-recovery/

Zhang, L., Zhang, D., Fang, J., Wan, Y., Tao, F., & Sun, Y. (2020). Assessment of mental health of Chinese primary school students before and after school closing and opening during the COVID-19 pandemic. *JAMA Network Open, 3*(9), e2021482.

Zhao, Y., & Watterson, J. (2021). The changes we need: Education post COVID-19. *Journal of Educational Change, 22*(3), 3–12. https://doi.org/10.1007/s10833-021-09417-3

CHAPTER 4

Impacts on School-Based Learning

Abstract "Learning from home" is emblematic of the COVID-19 pandemic. In March 2020, on the basis of public health advice, many educational systems suddenly shifted to learning at a distance, and students and teachers alike had little time to transition to this new phenomenon. The experience of learning from home has been interlaced with anxiety about the future of learning and schools and ongoing interruptions to face-to-face learning.

This abrupt transition revealed the disparities in educational and social capital experienced by children and young people around the world. Social inequities, like housing context, familial challenges, and socio-economic disadvantage, make learning from home difficult, if not impossible for many students. Drawing on international literature and our own research in Australia, this chapter explains how, for students experiencing disadvantage, there is an increased likelihood of learning loss resulting from the period of learning from home, which may impact future earnings and socio-economic security.

Learning from home has not only highlighted pre-existing educational inequities, but risks widening the achievement gap. Bourdieu's theories of capital are useful to understanding the underlying causes for variations of work done by students learning from home and the impetus to foreground equity in building the resilience of education systems to respond to crises.

E. S. Rudling et al., *Education and Equity in Times of Crisis*,
https://doi.org/10.1007/978-3-031-18671-4_4

Keywords COVID-19 pandemic • Learning from home • Overcrowded housing • Insecure housing • Material basics • Learning outcomes • Access to education

Introduction

Social inequalities in learning outcomes are well-established internationally. Students from disadvantaged backgrounds—especially those living in poverty—tend to have worse academic results than their more privileged peers (Lamb et al., 2020; Stevens & Dworkin, 2019; Teo, 2018; Thompson, 2019; Volante et al., 2019).

Of particular concern is that students who are missing out on early educational milestones are likely to continue to be behind—and even fall further behind—as they progress through school (Heckman & Mosso, 2014; Lamb et al., 2020). Quality early childhood education and care is especially critical for children from disadvantaged families, both to support their development and prepare them for school, but these children are less likely to attend preschool (Warren et al., 2016).

Regardless of student effort or capabilities, privilege confers benefits to support learning that disadvantaged families simply do not have access to, including material resources and a space that is conducive for study in the home. Theories of capital (Bourdieu, 1984) explain processes by which the early immersion in resource-rich environments provide children with a solid foundation for success at school. In addition to social and cultural capital providing children with access to materials for learning, play and creative expression, it facilitates a home life with the infrastructure that fosters successful learning. The disparity between capital-rich and capital-poor home environments is thrown into even greater relief during times of crisis.

In each of the sections below, we first provide an overview of equity concerns pre-pandemic in relation to the section topic, before examining how each played out during the COVID-19 pandemic.

School Learning from Home

During the COVID-19 pandemic, millions of school students suddenly had to engage with "school learning" from their home. In public discourse, this was sometimes referred to as "home-schooling" or "distance education". Both are misnomers, due to differences between these well-established options and "learning from home" during COVID-19.

Home-schooling is a legally recognised alternative to enrolling in school for students of compulsory school age in Australia and many other countries. Numbers of home-schooled students are estimated at more than two million in the United States, 80,000 in England, and 22,000 in Australia (Slater et al., 2020). Parents/carers who want to home-school their child must apply for registration with the relevant authority. In Australia this varies between states and territories[1]. In the case of home schooling, the parent/carer has actively chosen to be responsible for the child's education (English, 2021; Jolly & Matthews, 2020; Slater et al., 2020). This undertaking requires commitment and planning, as well as accountability to the registration authority. As a result, the parent/carer typically prepares an education program for their child that makes use of a wide range of learning resources. Commonly, the parent/carer who is the designated educator for home schooling is relatively highly educated (Hattie, 2020, Slater et al., 2020). School students enrolled through distance education are located at home for most of their learning but—unlike home-schooling—they are enrolled through a distance education provider, such as School of the Air in Australia. Its main use is for students living in geographically remote or isolated areas. Parents/carers provide supervision and a study space with relevant resources (including a computer and internet access), but the provider is responsible for the actual program of learning (Crump & Twyford, 2010; Halsey, 2018).

A Markedly Different Experience

The experience of "learning from home" during the COVID-19 pandemic was markedly different to both home-schooling and distance education. First, the move to learning from home during the pandemic was not a choice. It was imposed on students, teachers, and parents/carers alike. English (2021, p. 38) argues that some parents/carers turn to home-schooling "by accident not design" and "in response to a risk their

[1] See NSW Education Standards Authority, home-schooling registration (https://educationstandards.nsw.edu.au/wps/portal/nesa/regulation/home-schooling/home-schooling-registration); Queensland Department of Education, home education website (https://education.qld.gov.au/schools-educators/other-education/home-education); Victorian Registration and Qualifications Authority, home education website (https://www.vrqa.vic.gov.au/home/Pages/homeeducation.aspx).

child has experienced in school". However, during the pandemic there was not even an "accidental" choice—and the risk was not perceived by parents or carers in relation to their individual child, but by governments in relation to all children and adults. There was therefore a sense of powerlessness for families.

Secondly, there was, at least in the first instance, little time to plan and prepare—both by parents/carers and by teachers and education systems. Families did not necessarily have a suitable space set up for learning (also see below)—especially if they had several children—and if adults had to perform their job from home. They also did not have a thorough learning plan, prepared by the parent/carer as home educator or by the distance education provider. Having to use unfamiliar online learning platforms (such as Zoom, Seesaw, Canvas, Microsoft Teams) was an issue for some schools, teachers, students, and families. A school-based participant in our research during the first lockdown in 2020 explained:

> At this primary school the platforms that are now being used were not previously used and there was little opportunity to train the students in how to use them. If there was 3-4 weeks to get this up and running this would have been less of an issue, as students would be familiar. These are also not the platforms that are used by the parents—the majority of parents are unfamiliar with them and find it difficult to assist the students. There is a longer transition period because of this. (School based staff as cited in Brown et al., 2020, p. 28)

In contrast, in Singapore the online Student Learning Space[2] had been rolled out to all schools since 2017 as part of a focus on developing twenty-first century competencies and therefore many children were already used to that platform (Lau, 2020), although inequalities in access remained (Ang, 2020; Lee & Yeo, 2020).

Schools across the world did their best to set up contact time (usually via videoconferencing) and activities students could do independently. Many agencies (government departments of education, non-governmental organisations, media organisations) also quickly generated learning resources (often online) and tips. Nevertheless, there were concerns about the quality of learning.

[2] See https://www.moe.gov.sg/education-in-sg/student-learning-space.

The comment from 7-year-old Harriet in an Australian newspaper article on students' experiences of learning from home during the pandemic is poignant: "My mum is working so my dad helps me when I need it, but I find he much better suits his normal job as a fireman compared to teaching" (as cited in Chrysanthos, 2020, para. 4). Of course, nobody would expect teachers to suddenly be great fire fighters in a bushfire emergency. Nevertheless, Harriet's comment highlights the difficulty of parents/carers being expected to support their children's learning from home. This is especially alarming for children in disadvantaged families. As Hattie (2020, para. 12) pointed out in relation to the difference between home-schooling and learning from home during the pandemic: "the takeaway is that only parents who are well educated and who have attributes of good teachers are likely to home school with some success".

Third, the experience of learning from home during the pandemic has been woven through with anxiety—not only about health, but also about when and how learning would shift from home back to school. In home-schooling and distance education, that decision usually (although not always) is taken by parents/carers and students. During the COVID-19 pandemic, uncertainty has reigned. In many locations around the world there were frequent changes in policy—in response to changes in infection rate and health protocols—about which students could attend school on some, all, or no days of the week.

The United Nations Educational, Scientific and Cultural Organization (UNESCO, 2021a) provides an interactive map highlighting changes in full and partial school closures globally, beginning on 17 February 2020. For example, leaving out regular school holidays, the United Kingdom moved from schools being fully open to partial closure on 28 February 2020, to full closure on 20 March, partial closure on 1 June, fully open on 1 September, full closure on 5 January 2021, partial closure on 22 February 2021, and fully open on 8 March. In specific regions, changes may have been even more frequent.

In Australia, there has been a pattern of so-called "snap lockdowns", which are implemented very quickly, with little notice to the community. As Russell et al. (2021, para. 21) argue: "a snap lockdown can be very distressing for children and adolescents, many of whom are still struggling, exacerbated by the very difficult process of managing uncertainty". An education system officer in our research noted that six months after schools were re-opened "We've still got a lot of students who haven't yet returned to education". He attributed this to "paranoia and misinformation" and emphasised the need "to get that message out there and build confidence to come back to school". The crucial role of clear communication and the risk of misinformation have been widely recognised in the context of the pandemic (Donovan, 2020).

Learning Spaces at Home

For school students living in insecure or overcrowded housing, learning was already compromised before the pandemic due to disrupted sleep, lack of space to study, and the impact of noise levels on concentration. Such living conditions have been shown to have a negative effect on attendance, suspension and exclusion, and academic achievement (Harker, 2006). In terms of insecure housing, the Australia's Children report (Australian Institute of Health and Welfare [AIHW], 2020) indicates that, pre COVID-19, more than one-fifth of children aged 0–14 lived in households experiencing housing stress, with one-parent households being under the most stress. Family mobility (frequently moving home) is more common among families of low socioeconomic backgrounds and students in low socioeconomic status (SES) schools and tends to be in reaction to circumstances such as chasing work, a lease not being renewed, or family breakdown (Sorin & Iloste, 2006; Taylor & Edwards, 2012). Moving house often means also moving school, and therefore having to adjust to a different teacher, learning program, and peers. School mobility has been linked with reductions in attendance, engagement with learning, academic achievement, wellbeing, and mental health (Fantuzzo et al., 2012; Redmond & Skattebol, 2014; Skattebol et al., 2012; Sorin & Iloste, 2006; Taylor & Edwards, 2012; Winsper et al., 2016).

Living in overcrowded housing is also much more common for children in low socioeconomic areas, partly due to the high cost of housing in many parts of Australia (AIHW, 2020). Overcrowded housing can lead to disrupted sleep and ability to concentrate, which in turn can cause emotional and behavioural problems as well as reduced school performance. Moreover, in overcrowded homes students are less likely to have access to a quiet study space to do homework or prepare for exams; and space to play (Harker, 2006; Rice, 2006; Skattebol et al., 2012; Solari & Mare, 2012). Highly vulnerable children and young people are more likely to live in homes that are overcrowded. Two UK reports highlight that while students living in bad and overcrowded housing tend to consider school as very important, their actual experiences of school are far more negative than for those not living in overcrowded homes (Harker, 2006; Rice, 2006). They had lower attendance, were more often excluded from school, more likely to be held back a grade, and more likely to leave school without a credential.

Learning Spaces During the Pandemic

In the COVID-19 context, the negative impacts of insecure and overcrowded housing are likely to be exacerbated. A statutory office holder in our research noted:

> We know that play opportunities are really restricted by the type of housing you live in as well. The thing with a lot of the more vulnerable kids is they're not in their own homes, they're in rentals… All of these things just go to the point that any crisis just exacerbates the inequality that's already existing. (Advocacy Groups and Services officer)

During periods of lockdown, when children and adults are required by governments to be at home for extended periods, inadequate spaces for learning at home are a particular concern. The impact of inadequate spaces for doing schoolwork has been identified in previous studies, for example, Tessa (as cited in Skattebol et al., 2012, p. 85) used a homework centre in her school because, "It is too hard to do it in the house; too much kids here". The opportunity of accessing such extracurricular support largely ceased during lockdowns.

In our research, over 90% of survey respondents considered "adequate space" for learning to be a critical issue for learning at home for vulnerable students (Brown et al., 2020). Key informants also pointed to overcrowding as a factor making learning at home problematic.

> The other thing is that there's often some pretty crowded living arrangements as well for lots of families. They live in small places, with lots of people. So actually, being able to be in a place where's no other room that people can go to. There's generally bedrooms that are shared and then a living space. So I think that makes it really kind of tricky. (Advocacy Groups and Services officer)

> A lot of houses have no places for children to be able to study, or no separate places. So, they're often in the way or in the midst of the general chaos. (Advocacy Groups and Services staff member)

In Singapore, often considered a country with strong access to technology and high educational achievement, concerns about inequality stemming from adequate learning spaces at home were expressed early on, including that: "… a child who has his own room may find it easier to concentrate on HBL [home-based learning], while three children sharing one laptop in the living room may have more difficulty focusing on the assignments" (Lee & Yeo, 2020, para. 11).

For students, regardless of whether they had their own study space at home or not, the physical learning environment at home during the pandemic did not tend to measure up to spaces at school. School students wrote in an Australian newspaper story featuring their stories:

> I'm transforming into an indolent, fatigued pig, sitting down for six hours straight and staring at my computer screen until I have accomplished my assignments. (Angelina, age 14, as cited in Chrysanthos, 2020, para. 37)

> Our activities ended, our sports and bands,
> Headphones replaced our outstretched hands,
> Our classrooms substituted for virtual screens,
> Our days spent, watery eyes glued to teams,
> Our lunchtimes enjoyed, indoors, and alone,
> Our only interaction through the screen of a phone. (Excerpt from poem by Ben, age 15, as cited in Chrysanthos 2020, para. 49)

The same newspaper story included a poignant drawing by a Year 9 student Hannah Lemon, with washed out colours indoors, bright sun and a rainbow outside, and prison-type bars in the window frame (Chrysanthos, 2020). Her drawing indicates that despite having a chair, desk, bookcase and plants, her space for learning at home felt like prison.

Advice on setting up a learning space has been provided by various agencies as part of the information for parents/carers about supporting their children to learn from home during lockdowns. For example, the education department in the state of Queensland in Australia offered the following advice:

> Create a quiet and comfortable learning space. A space for extended learning should be a family space, rather than a bedroom. Your children may have a regular place for doing homework under normal circumstances, but this space may not be suitable for working in for an extended period of time. It

should be a place that can be quiet at times and if possible, have internet access. (Queensland Department of Education, 2020)

In overcrowded homes, these suggestions may simply be impossible. As noted above, often these homes do not have "a regular place for doing homework under normal circumstances", let alone meet the additional criteria in the Department's advice. This highlights that adequate space for learning is a critical equity issue for families in homes that are already stretched and ill-equipped for learning.

Learning Inequalities

As noted in the introduction to this chapter, inequality in learning outcomes is well established between students from low and high SES groups in countries around the world and between First Nations and non-Indigenous students in countries with colonial heritage. In the United Kingdom (Crawford et al., 2014; Goodman & Gregg, 2010; Shaw et al., 2017) and North America (Janus & Duku, 2007; Kuhfeld et al., 2018) children from low SES backgrounds commence school behind their high SES peers, and the gap widens further over time.

In Australia, based on data from the Organisation for Economic Co-operation and Development (OECD) Programme for International Student Assessment for science and mathematics, secondary school students from low SES backgrounds in low SES schools are about three years behind their high SES peers in a high SES school (McConney & Perry, 2010; Perry & Lubienski, 2014). On the 2018 NAPLAN test[3], in Year 3 (primary school) about 15% of Indigenous students in Australia scored below the national minimum standard on reading and on numeracy, compared to 1–2% of non-Indigenous students. Conversely, only about 7% of Indigenous students scored in the top band on reading, compared to over 30% of non-Indigenous students (Department of the Prime Minister and Cabinet, 2020). According to Bourdieu's (1984) theory of capital—dominant culture sets the "rules of conduct" (p. 331) by ascribing value to the knowledges and knowledge systems of dominant culture. In so doing, this constructs Indigenous knowledges and ways of knowing as outsider/lesser, excluding/devaluing them in standardised testing regimes. This is a

[3] NAPLAN is the National Assessment Program Literacy and Numeracy, administered in Australian schools in Years 3, 5, 7 and 9.

part of the machinery of educational disadvantage and exclusion in Australia, as in other countries where education systems favour colonised knowledge.

Figure 4.1 shows findings from a major analysis of educational opportunity in Australia (Lamb et al., 2020), highlighting disadvantage for low SES and First Nations students in the early years, start of secondary education, and completion of secondary education.

This data demonstrates the achievement gap that exists between low and high SES students and between Indigenous and non-Indigenous students, clearly rendering that the landscape of inequality is directly connected with education outcomes.

In relation to the experience of education during COVID-19 interruptions to schooling, research on the effect of the holiday break on student achievement is relevant. Established in research initially conducted in the United States (Heyns, 1978), the phenomenon of the so-called "summer learning loss" is used to explain how, during the summer holiday, the learning achievements of students from low SES backgrounds tend to go backwards, while those of high SES students increase. Research in the Australian state of Tasmania by Tomaszewski et al. (2022) shows attendance rates for low SES students worsened as a result of the pandemic, which is expected to adversely affect educational outcomes. The cumulative effect of long holiday periods may, at least partially, account for lower educational attainment of children from low-income families, those with learning difficulties and with English as a second language (Stewart et al., 2018; Van Lancker & Parolin, 2020).

This "summer learning loss" has been demonstrated most clearly in the United States, where at 12 weeks the summer break is much longer than in Australia, New Zealand, and European countries. In the latter countries research shows less dramatic summer learning loss and/or demonstrates that disadvantaged students catch up this loss once school resumes (Hattie 2020; Shinwell & Defeyter, 2017). In contrast, in the United States, summer learning loss may work to further widen the gap between low and high SES students which already exists at the start of schooling (Alexander et al., 2007).

Learning Experiences During the Pandemic

Although the phrase "school closures" is frequently used as a short-hand during the COVID-19 pandemic, in fact school support for learning

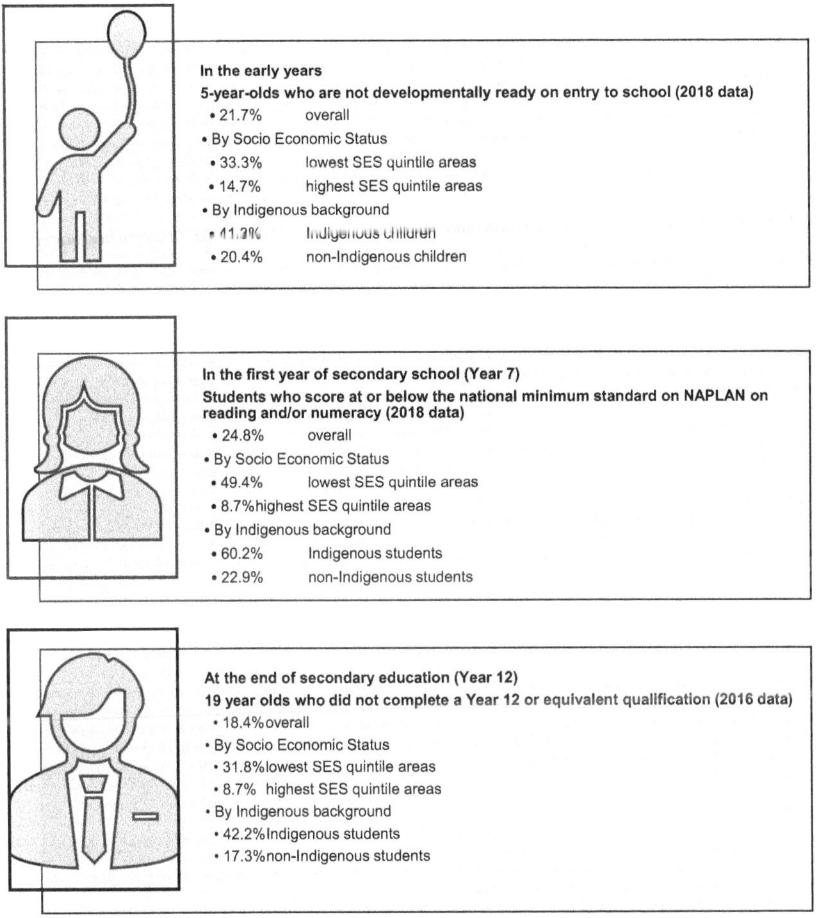

Fig. 4.1 Educational opportunity in Australia. Data from Lamb et al. (2020)

continued. For most students, this meant the provision of learning materials usually supplemented with online face-to-face teaching. In addition, some students continued to attend school in person, including the children of key workers and children deemed vulnerable (Bayrakdar & Guveli, 2020; Brown et al., 2020; Lee & Yeo, 2020). Nevertheless, the turn to remote learning meant a reduction in instructional time as well as learning time.

While comparisons have been made with "summer learning loss" (see above), a major difference is that during the pandemic students have been away from school during actual school time, rather than during a holiday. Moreover, in many countries, reduced access to learning at school extended far longer than even the lengthy American school holiday. The UNESCO (2021b) global dataset on the duration of school closures shows that between March 2020–June 2021, 30 countries had 30 or more weeks of full closure (i.e., all schools nation-wide) and an additional 30 (different) countries had 30 or more weeks of partial closure (i.e., in some parts of the nation, for certain grades, or a reduction of in-person teaching). In total 110 countries experienced more than 30 weeks of full and partial school closures combined.

To illustrate, the Netherlands had 12 weeks of full school closures, and 19 weeks of partial closures (UNESCO, 2021b). In primary and secondary schools in the Netherlands students were unable to attend school not only due to repeated general lockdowns (nationally or in a region), but also to enable schools to implement physical distancing (with different students attending different days), due to a lack of school staff, and to manage site-specific COVID-19 infection (Inspectie van het Onderwijs, 2020a, 2020b). For example, at the start of the 2020–2021 37% of secondary schools had needed to temporarily require some classes of students to learn from home, even though there was no general lockdown.

Disparities in Extent of Learning While Learning from Home
Respondents to our survey during the first lockdown indicated that time spent on learning at home was variable, although 2-3 hours a day was a common estimate (Brown et al., 2020). Similarly, an online survey of parents in the Czech Republic found children generally spent 2–4 hours studying during learning from home (Brom et al., 2020). A parent and an educator in our research shed light on the variety of experiences:

> I have four children and it varies enormously depending on the individual child's ability to self-motivate and engage. For example: My 14-year-old is reasonably self-motivated but frustrated that he is not learning new concepts, the teachers are mainly setting revision ... This compares to my 5-year-old who is not the least bit interested in online learning and likes to sit on my lap while I try to work. (Parent/educator as cited in Brown et al., 2020, p. 28)

While that was one parent's account from a busy household, an educator offered another view which revealed the haphazard and variable amount of learning that her class of students engaged in.

> Six are completing no work at all (as admitted by parents and students); several are completing only the final activities and not the teaching activities that are leading up to it; three are focusing on topics the student wants to learn about (not curriculum-linked, just topics of interest). (School-based staff as cited in Brown et al., 2020, p. 30)

According to Forster (2020) learning was also reduced because Australian schools prioritised wellbeing during the pandemic over the completion of schoolwork. This prioritisation of wellbeing extended to teachers, with one school-based leader stating that staff wellbeing came first "even to the point where we sent the kids home for the first week with no learning" (as cited in Forster, 2020, para. 14).

In the United States, analysis of a large data set on the use of an online mathematics program before and after students shifted to learning from home in March 2020 showed that by late April 2020 students' learning progress had decreased by half in classrooms in low SES postcode areas but had not decreased in classrooms in high SES areas (Goldstein, 2020).

In Australia, an estimate is that the gap in learning achievement between disadvantaged and other students increased three times faster during the first lockdown than normally, and that disadvantaged students will have lost 1–2 months' worth of learning during two months of remote learning (Sonnemann & Goss, 2020). This loss adds to pre-existing inequalities. For example, the same report indicates a gap of 16 months in reading and numeracy between disadvantaged students and all other students in Year 5, prior to the pandemic (Sonnemann & Goss, 2020). This gap has widened and at a faster rate due to interruptions to schooling brought about by the COVID-19 pandemic.

Remarkably, some highly vulnerable students may have done more work at home than at school; for example, students enrolled in "flexible learning programs" offering a second chance for completing school for young people for whom mainstream school has not worked well. These programs are characterised by changing the way they offer education to suit each student (te Riele, 2014). A Department of Education staff member in Australia noted:

> [At] one of our sites we had a 90% return [of completed work] for two weeks. For each pickup [of hard copy schoolwork], 90% of the kids responding and having completed all the work. But it was very personalised to the students. ... We are talking about those young people who have been disengaged from learning for a long time so we're pretty impressed. (Education system staff as cited in Brown et al., 2020, p. 30)

Students from flexi-schools in other Australian research commented that the high level of support from their school continued during COVID-19 school closures (McDaid et al., 2021). Moreover, they were already used to individualised and self-paced learning, which meant they felt well-prepared when they had to learn from home during lockdowns: "The best thing is we just do it at our own pace. Everyone does their own work. We finish when we finish as long as, you know, we try and get it right. [...] Just do your work" (Flexi-school student as cited in McDaid et al., 2021, p. 32).

During learning from home, formal learning could be supplemented by an increase in informal learning since children spent more time with their families. The ability of families to provide such experiences relies on access to economic and cultural capitals which affords access to enriched and enriching experiences of learning. For example, in Singapore a parent (with a Master's degree and her own home-based business) purchased a subscription to an online resource offering:

> ... bite-size summaries by the authors of books on topics such as entrepreneurship and self-development. "After they watch it, I want them to come back and report to me, and tell me what they have learnt, so that I can also learn from them," [the mother] says. Dylan [her son], for example, shared tips he learnt about improving one's Instagram bio, which she promptly used. (Lee & Yeo, 2020, para. 42)

In Spain, in both primary and secondary education children of more highly educated parents participated more often in after-school activities than children of parents who had only completed compulsory education—and they were more likely to continue these activities during the lockdown (Bonal & González, 2020). This was due in part to the nature of the activities (art, music, foreign languages versus sport) and in part because lower-educated parents were more likely to voluntarily relinquish activities that could have continued online "due to difficulties in coping with the costs during times of crisis" (Bonal & González, 2020, p. 650). In Australia, when after school activities resumed, some students chose not to re-engage with these because they were focused on catching up on their schoolwork: "I didn't really want to play [sport] this year. It was just too hard. [...] Because of COVID and also like to try and get my grades back up" (Secondary school student as cited in McDaid et al., 2021, p. 33).

For students who continued to attend school, time spent on learning also varied. It was not uncommon, at least in the first experiences of lockdown, for schools to provide students who were on-site with supervision to undertake the same online learning as students were doing school at home (Australian Capital Territory Education, 2020; Brown et al., 2020). Tomaszewski et al. (2022) tracked pandemic school attendance rates for students in the upper years of secondary school in Tasmania once students returned to on-site learning, with lower SES students more likely to have reduced attendance during this time, in turn likely contributing to learning loss.

In the next section we examine the impact of the pandemic specifically on foundational learning during primary education.

Foundational Learning

It is well established that developmental and wellbeing benefits arise from children's participation in quality early childhood education and care. Increasing participation in early childhood education and care is a joint focus of international organisations such as the United Nations' OECD, the World Bank, and the International Labour Organisation. Nobel Prize recipient James Heckman (2007) has influenced an emerging consensus among policymakers of the economic and social policy sense of prioritising access to quality early childhood education and care experiences—particularly for children who are regarded as "vulnerable".

Crucially, economic analyses have shaped a policy framing about whether early childhood education and services are funded universally or if financial support for provision is targeted towards children and families who may are regarded as vulnerable. Unlike education for all in the compulsory years of schooling, early years opportunities are subject to economic rationing (Shelley, 2021). In Australia, Council of Australian Governments (CoAG) early years policy texts describe a fair and just society and a right to equitable access and participation (CoAG, 2009a, 2009b, 2009c). Yet, inequality of access and participation is a dominant feature of Australia's early childhood education and care system. During the early stages of the pandemic in 2020, many families had withdrawn their children from care. Some families were motivated by health concerns and some because they were providing care to their children at home as a result of being stood down, losing employment, or working from home. Early Childhood Australia, a peak body, estimated that about 640—or 4%—of early childhood education and care centres closed (Hurst, 2020). For a short period (April—June 2020) Australia benefitted from a radical policy shift to support access and equity. In April 2020, the Government announced new funding arrangements for the early childhood education and care sector to make childcare services fee-free for families (Klapdor, 2020).

Primary education serves many purposes. In relation to academic learning, one of those purposes is to ensure all students master the foundational learning needed for ongoing academic success. For young students, this relies strongly on play-based pedagogies. Inequalities in access to play-based learning were evident during the pandemic. A survey of over 35,000 families in Spain early in the pandemic (when schools had been largely closed for two weeks, at the end of March 2020) indicated acute differences for 3–8-year-old children (Bonal & González, 2020):

- In families with a highly educated adult, it was far more common for a child to read with an adult every day; to practice a foreign language several times a week; and to engage in sport several times a week.
- In families where adults had only completed compulsory education, it was far more common for children to play video games several times a week; and watch educational television programs every day.

Bonal and González (2020) conclude that:

> In such an extreme situation of school absence, it appears that families with less cultural capital made more use of external resources to support their children's learning activities, while families with more cultural capital were more confident in their own abilities to respond to their children's learning needs. (p. 649)

When a broader lens is applied, beyond learning associated predominantly with curriculum, a more positive picture emerges. Research conducted with primary school aged children in New Zealand and in the Australian state of Tasmania has showed significant variation between children's experiences and highlighted the complexity of learning that each child experienced (Bourke et al., 2021). Importantly this work reveals that alone and with the support of family children can and do create the necessary conditions for learning in challenging times. There is a significant body of scholarly and mass media commentary focused on what primary school aged students are missing out on. Yet, interestingly, this research points to children's own largely positive accounts of their informal learning.

Van de Werfhorst (2021) argues that "Impacts of school closures [during the pandemic] on inequalities in learning [...] may be most prominent in primary education, because the foundation is laid for future learning progression" (p. 2). For example, literacy is an essential and lifelong enabler for engaging fully in education, society, employment, and civic life (Stewart et al., 2019). Unsurprisingly, in the context of the impact of the pandemic on learning, literacy has been raised as a concern by educators:

> Literacy, that would be my biggest worry, that we'll lose ground here ... We have to make sure we have the literacy things in place because that's one of the things I'm concerned about. If literacy is the only thing we keep progressing that is number one, because we know that is the number one predictor of success. It is critical that we can get on top of how we can support ongoing literacy development and that's pretty hard when they're at a distance. (School-based staff as cited in Brown et al., 2020, p. 30)

The Netherlands has, as Van de Werfhorst (2021, p. 2) calls it, a "rich Dutch data infrastructure" with standardised test results enabling comparison between groups. The National Cohort Study Education (Nationaal Cohortonderzoek Onderwijs, 2021) conducted analysis of Dutch primary school students' test scores, comparing learning growth over a year of the pandemic with learning growth in the two preceding years, based on a set of tests conducted annually in February/March. As noted above, there has been variety in partial school closures within countries (see UNESCO, 2021a, 2021b). The results demonstrate an overall reduction in learning growth during 2020 of 19% in reading comprehension, 7% in spelling, and 9% in mathematics (Haelermans et al., 2021a, p. 1). Importantly the findings also demonstrate implications for equity (Haelermans et al., 2021b, 2021c), with more severe learning delays for:

- Students from low socio-economic backgrounds, in spelling and mathematics;
- Students from single parent families, in reading comprehension and mathematics;
- Students from large families (three or more children living at home), in spelling and mathematics;
- Schools with relatively many students from low socio-economic backgrounds;
- Schools in non-urban areas;
- Small schools.

Similarly in the United Kingdom (UK), the impact of the pandemic on foundational learning has been most significant for disadvantaged students, who "lost, on average, approximately 2.2 months" in reading (half a month more than non-disadvantaged students) and "around 4.5 months" in mathematics (a month more than non-disadvantaged students) (Renaissance Learning and Education Policy Institute, 2021, pp. 18-19). The report argues that based on the UK Spring term 2021, "the extent to which disadvantaged pupils lost learning…appears to be equivalent to undoing two-thirds of the progress made in the last decade on closing the gap in disadvantaged primary schools" (Renaissance Learning and Education Policy Institute, 2021, p. 36).

Conclusion

This chapter demonstrates how existing inequalities in learning have been exacerbated by the COVID-19 pandemic, drawing not only on our own research in Australia but on experiences and data from a wide range of countries internationally. Shining through all this is the deep concern on the ground (by students, parents/carers, and educators) as well as in research and policy about both the immediate repercussions for vulnerable students and the long-term repercussions of a growing gap in formal learning outcomes between more and less advantaged groups. The focus of this chapter and this book is mostly on equity implications within developed countries. We recognise, however, the considerable gaps also between more and less advantaged countries (United Nations Educational, Scientific and Cultural Organization, United Nations Children's Fund, the World Bank, and Organisation for Economic Co-operation and Development, 2021).

The notion of the forms of capital available to students and families has proven to be of great relevance. For example, in terms of material capital, learning spaces are especially vital. Students living in small or overcrowded homes, without access to a dedicated and quiet study space, have been particularly disadvantaged. The difference between traditional home-schooling and learning from home during the pandemic highlights the importance of parents'/carers' cultural capital for supporting their child's learning in the home environment.

References

Alexander, K. L., Entwisle, D. R., & Olson, L. S. (2007). Lasting consequences of the summer learning gap. *American Sociological Review, 72*(2), 167–180. https://doi.org/10.1177/000312240707200202

Ang, J. (2020, April 8). Parents struggle to juggle working from home and supervising kids' home-based learning amid Covid-19 crisis. *The Straits Times*. Retrieved July 5, 2021, from https://www.straitstimes.com/singapore/juggling-working-from-home-and-supervising-kids-home-based-learning-is-a-struggle-for-some

Australian Capital Territory Education. (2020). *Home learning resources for students and families: Online tools to support learning at home*. Retrieved October 20, 2021, from https://sites.google.com/ed.act.edu.au/act-home-learning/family-guides/online-learning

Australian Institute of Health and Welfare. (2020). *Australia's children* (Cat. No. CWS 69). Retrieved April 4, 2021, from https://www.aihw.gov.au/getmedia/6af928d6-692e-4449-b915-cf2ca946982f/aihw-cws-69-print-report.pdf.aspx?inline=true

Bayrakdar, S., & Guveli, A. (2020). *Inequalities in home learning and schools' provision of distance teaching during school closure of COVID-19 lockdown in the UK* (ISER Working Paper Series, No. 2020-09). Institute for Social and Economic Research (ISER), University of Essex. Retrieved August 15, 2020, from https://www.econstor.eu/bitstream/10419/227790/1/1703719352.pdf

Bonal, X., & González, S. (2020). The impact of lockdown on the learning gap: Family and school divisions in times of crisis. *International Review of Education, 66*(5), 635–655. https://doi.org/10.1007/s11159-020-09860-z

Bourdieu, P. (1984). *Distinction: A social critique of the judgement of taste* (R. Nice, Trans.). Harvard University Press.

Bourke, R., O'Neill, J., McDowall, S., Dacre, M., Mincher, N., Narayanan, V., Overbye, S., & Tuifagalele, R. (2021). *Children's informal learning at home during COVID-19 lockdown*. New Zealand Council for Educational Research and Massey University. Retrieved September 23, 2021, from https://www.nzcer.org.nz/system/files/NZC_Children%27s%20informal%20learning%20Final%2023-8-21.pdf

Brom, C., Lukavský, J., Greger, D., Hannemann, T., Straková, J., & Švaříček, R. (2020). Mandatory home education during the COVID-19 lockdown in the Czech Republic: A rapid survey of 1st–9th Graders' parents. *Frontiers in Education*. https://doi.org/10.3389/feduc.2020.00103

Brown, N., te Riele, K., Shelley, B., & Woodroffe, J. (2020). *Learning at home during COVID-19: Effects on vulnerable young Australians* (Independent Rapid Response Report). Peter Underwood Centre for Educational Attainment, University of Tasmania. Retrieved May 26, 2020, from https://www.utas.edu.au/__data/assets/pdf_file/0008/1324268/Learning-at-home-during-COVID-19-updated.pdf

Chrysanthos, N. (2020, April 10). I never thought I'd say it, but I miss going to school. *The Sydney Morning Herald*. Retrieved August 18, 2020, from https://www.smh.com.au/national/nsw/i-never-thought-i-d-say-it-but-i-miss-going-to-school-20200409-p54ik5.html

Council of Australian Governments. (2009a). *Closing the gap: National partnership agreement on indigenous early childhood*. Australian Government.

Council of Australian Governments. (2009b). *Investing in the early years: A national early childhood development strategy*. Australian Government.

Council of Australian Governments. (2009c). *The national quality standard for early childhood education and care*. Australian Government.

Crawford, C., Macmillan, L., & Vignoles, A. (2014). *Progress made by high-attaining children from disadvantaged backgrounds*. Social Mobility & Child Poverty Commission, Department of Education, England. Retrieved February 10, 2021, from https://www.gov.uk/government/uploads/system/uploads/attachment_data/file/324501/High_attainers_progress_report_final.pdf

Crump, S., & Twyford, K. (2010). Opening their eyes: E-learning for rural and isolated communities in Australia. In K. A. Schafft & A. Youngblood Jackson (Eds.), *Rural education for the twenty-first century: Identity, place, and community in a globalizing world* (pp. 253–274). Pennsylvania State University Press.

Department of the Prime Minister and Cabinet. (2020). *Closing the Gap report 2020*. Commonwealth of Australia, Department of the Prime Minister and Cabinet. Retrieved April 10, 2021, from https://ctgreport.niaa.gov.au/

Donovan, J. (2020). Concrete recommendations for cutting through misinformation during the COVID-19 pandemic. *American Journal of Public Health, 110*(3), S286–S287. https://doi.org/10.2105/AJPH.2020.305922

English, R. (2021). The accidental home educator: A new conceptualisation of home education choice. In R. English (Ed.), *Global perspectives on home education in the 21st century* (pp. 30–48). IGI Global.

Fantuzzo, J. W., LeBoeuf, W. A., Chen, C.-C., Rouse, H. L., & Culhane, D. P. (2012). The unique and combined effects of homelessness and school mobility on the educational outcomes of young children. *Educational Researcher, 41*(9), 393–402. https://doi.org/10.3102/0013189X12468210

Forster, D. (2020, November 9). Is learning more important than well-being? Teachers told us how COVID highlighted ethical dilemmas at school. *The Conversation*. Retrieved November 10, 2020, from https://theconversation.com/is-learning-more-important-than-well-being-teachers-told-us-how-covid-highlighted-ethical-dilemmas-at-school-144854

Goldstein, D. (2020, June 5). Research shows students falling months behind during virus disruptions. *The New York Times*. Retrieved June 7, 2020, from https://www.nytimes.com/2020/06/05/us/coronavirus-education-lost-learning.html

Goodman, A., & Gregg, P. (2010). *Poorer children's educational attainment: How important are attitudes and behaviour?* Joseph Rowntree Foundation. Retrieved March 11, 2021, from https://www.jrf.org.uk/report/poorer-children%E2%80%99s-educational-attainment-how-important-are-attitudes-and-behaviour

Haelermans, C., Van der Velden, R., Aarts, B., Bijlsma, I., Jacobs, M., Smeets, C., Van Vugt, L., & Van Wetten, S. (2021a). *Leergroei deels hersteld, behalve bij begrijpend lezen* [Learning growth partially recovered, except for reading comprehension] (NCO Factsheet No. 5). Nationaal Cohortonderzoek Onderwijs. Retrieved August 11, 2021, from https://www.nationaalcohortonderzoek.nl/factsheets/

Haelermans, C., Van der Velden, R., Aarts, B., Bijlsma, I., Jacobs, M., Smeets, C., Van Vugt, L., & Van Wetten, S. (2021b). *Kwetsbare leerlingen nog steeds het hardst geraakt* [Vulnerable students still most affected] (NCO Factsheet No. 6). Nationaal Cohortonderzoek Onderwijs. Retrieved August 11, 2021, from https://www.nationaalcohortonderzoek.nl/factsheets/

Haelermans, C., Van der Velden, R., Aarts, B., Bijlsma, I., Jacobs, M., Smeets, C., Van Vugt, L., & Van Wetten, S. (2021c). *Vertraging in leergroei verschilt tussen scholen* [Delay in learning growth varies between schools] (NCO Factsheet No. 7). Nationaal Cohortonderzoek Onderwijs. Retrieved August 11, 2021, from https://www.nationaalcohortonderzoek.nl/factsheets/

Halsey, J. (2018). *Independent review into regional rural and remote education—Final report*. Commonwealth of Australia. Retrieved July 22, 2021, from https://www.dese.gov.au/quality-schools-package/resources/independent-review-regional-rural-and-remote-education-final-report

Harker, L. (2006). *Chance of a lifetime: The impact of bad housing on children's lives*. Shelter. Retrieved July 22, 2021, from https://assets.ctfassets.net/6sxvmndnpn0s/4LTXp3mya7IigRmNG8x9KK/6922b5a4c6ea756ea94da71ebdc001a5/Chance_of_a_Lifetime.pdf

Hattie, J. (2020). *Visible learning effect sizes when schools are closed: What matters and what does not*. Retrieved August 11, 2021, from https://corwin-connect.com/2020/04/visible-learning-effect-sizes-when-schools-are-closed-what-matters-and-what-does-not/

Heckman, J. J. (2007). The economics, technology, and neuroscience of human capability formation. *Proceedings of the National Academy of Sciences, 104*(33), 13250–13255.

Heckman, J. J., & Mosso, S. (2014). *The economics of human development and social mobility* (Working Paper No. 19925). National Bureau of Economic Research. Retrieved February 18, 2021, from https://www.nber.org/papers/w19925

Heyns, B. (1978). *Summer learning and the effects of schooling*. Academic.

Hurst, D. (2020, April 2). Australian childcare operators fear they will have to close without a government lifeline. *The Guardian*. Retrieved September 9, 2021, from https://www.theguardian.com/australia-news/2020/apr/02/australian-childcare-operators-fear-they-will-have-to-close-without-a-government-lifeline

Inspectie van het Onderwijs. (2020a). COVID-19-monitor Inspectie van het Onderwijs meting 3—primair onderwijs [COVID-19-monitor education inspection measurement 3—primary education]. Retrieved March 5, 2021, from https://www.onderwijsinspectie.nl/onderwerpen/corona-onderzoeken/documenten/publicaties/2020/11/24/covid-19-monitor-po-derde-meting

Inspectie van het Onderwijs. (2020b). COVID-19-monitor Inspectie van het Onderwijs meting 3—voortgezet onderwijs [COVID-19-monitor Education Inspection measurement 3—secondary education]. Retrieved March 5, 2021, from https://www.onderwijsinspectie.nl/onderwerpen/corona-onderzoeken/documenten/publicaties/2020/11/24/covid-19-monitor-vo-derde-meting

Janus, M., & Duku, E. (2007). The school entry gap: Socioeconomic, family, and health factors associated with children's school readiness to learn. *Early Education and Development, 18*(3), 375–403. https://doi.org/10.1080/10409280701610796a

Jolly, J. L., & Matthews, M. S. (2020). The shifting landscape of the homeschooling continuum. *Educational Review, 72*(3), 269–280. https://doi.org/10.1080/00131911.2018.1552661

Klapdor, M. (2020). *COVID-19 economic response—Free childcare*. Parliament of Australia. Retrieved January 12, 2021, from https://www.aph.gov.au/About_Parliament/Parliamentary_Departments/Parliamentary_Library/FlagPost/2020/April/Coronavirus_response-Free_child_care

Kuhfeld, M., Gershoff, E., & Paschall, K. (2018). The development of racial/ethnic and socioeconomic achievement gaps during the school years. *Journal of Applied Developmental Psychology, 57*, 62–73. https://doi.org/10.1016/j.appdev.2018.07.001

Lamb, S., Huo, S., Walstab, A., Wade, A., Maire, Q., Doecke, E., Jackson, J., & Endekov, Z. (2020). *Educational opportunity in Australia 2020: Who succeeds and who misses out*. Centre for International Research on Education Systems, Victoria University, for the Mitchell Institute. Retrieved November 12, 2020, from https://www.vu.edu.au/sites/default/files/educational-opportunity-in-australia-2020.pdf

Lau, K. Y. (2020, August 12). Reforming education in the post-COVID-19 era. *EJ Insight*. Retrieved August 14, 2020, from https://www.ejinsight.com/eji/article/id/2548834/20200812-Reforming-education-in-the-post-COVID-19-era

Lee, V., & Yeo, S. (2020, April 18). How home-based learning shows up inequality in Singapore—A look at three homes. *The Straits Times*. Retrieved July 5, 2021, from https://www.straitstimes.com/lifestyle/how-home-based-learning-hbl-shows-up-inequality-in-singapore-a-look-at-three-homes

McConney, A., & Perry, L. B. (2010). Science and mathematics achievement in Australia: The role of school socioeconomic composition in educational equity and effectiveness. *International Journal of Science and Mathematics Education, 8*(3), 429–452. https://doi.org/10.1007/s10763-010-9197-4

McDaid, L., Povey, J., Tomaszewski, W., Cleary, A., Plage, S., Cook, S., Xiang, N., Zajac, T., Ridgway, T., & Western, M. (2021). *What do children, young people and families say about needs, and impact of COVID-19?* (Pillar Report No. 2). Institute for Social Science Research, University of Queensland. Retrieved June 11, 2021, from https://issr.uq.edu.au/files/21372/Ramsay%20Pillar%202_Final.pdf

Nationaal Cohortonderzoek Onderwijs. (2021). *Nationaal Cohortonderzoek Onderwijs* [National cohort study education]. Retrieved September 7, 2021, from https://www.nationaalcohortonderzoek.nl/

Perry, L. B., & Lubienski, C. (2014, March 13). Australian schools: Engines of inequality. *The Conversation*. Retrieved May 12, 2020, from https://theconversation.com/australian-schools-engines-of-inequality-23979

Queensland Department of Education. (2020). *Setting up a learning space*. Retrieved August 10, 2021, from https://education.qld.gov.au/curriculum/learning-at-home/wellbeing-of-students#setting-up

Redmond, G., & Skattebol, J. (2014). Filling in the details. Significant events and economic disadvantage among young people in Australia. *Zeitschrift für Pädagogik, 60*(4), 552–568. Retrieved April 14, 2020, from https://www.pedocs.de/volltexte/2017/14672/pdf/ZfPaed_2014_4_Redmond_Skattebol_Filling_in_the_details.pdf

Renaissance Learning and Education Policy Institute. (2021). *Understanding progress in the 2020/21 Academic Year. Complete findings from the Spring term*. Retrieved June 9, 2022, from https://assets.publishing.service.gov.uk/government/uploads/system/uploads/attachment_data/file/1062287/Understanding_progress_in_the_2020-21_academic_year_Complete_findings_from_spring_term_Oct2021.pdf

Rice, B. (2006). *Against the odds: An investigation comparing the lives of children on either side of Britain's housing divide*. Shelter. Retrieved April 12, 2020, from https://england.shelter.org.uk/professional_resources/policy_and_research/policy_library/against_the_odds

Russell, F., Koirala, A., Bowen, A., Danchin, M., & Goldfeld, A. (2021, March 31). We can't close schools every time there's a COVID outbreak. Our traffic light system shows what to do instead. *The Conversation*. Retrieved April 4, 2021, from https://theconversation.com/we-cant-close-schools-every-time-theres-a-covid-outbreak-our-traffic-light-system-shows-what-to-do-instead-158214

Shaw, B., Baars, S., Menzies L., Parameshwaran, M., & Allen, R. (2017). *Low income pupils' progress at secondary school*. Social Mobility Commission. Retrieved May 5, 2021, from https://www.gov.uk/government/uploads/system/uploads/attachment_data/file/594363/Progress_at_Secondary_School_report_final.pdf

Shelley, B. (2021). System reform in the early childhood education and care sector in Australia. In W. Pink (Ed.), *Oxford encyclopedia of school reform*. Oxford University Press. https://doi.org/10.1093/acrefore/9780190264093.013.1668

Shinwell, J., & Defeyter, M. A. (2017). *Investigation into summer learning loss in the UK. Implications for holiday club provision*. Frontiers in Public Health. https://doi.org/10.3389/fpubh.2017.00270

Skattebol, J., Saunders, P., Redmond, G., Bedford, M., & Cass, B. (2012). *Making a difference: Building on young people's experiences of economic adversity*. Social Policy Research Centre, University of New South Wales. Retrieved March 19, 2021, from https://www.thesmithfamily.com.au/-/media/files/research/reports/making-a-difference-2012.pdf?la=en&hash=B06CFAFA7AC2742346359EE3B5AE6418

Slater, E. V., Burton, K., & McKillop, D. (2020). Reasons for home educating in Australia: Who and why? *Educational Review*. https://doi.org/10.1080/00131911.2020.1728232

Solari, C. D., & Mare, R. D. (2012). Housing crowding effects on children's well-being. *Social Science Research, 41*(2), 464–476. https://doi.org/10.1016/j.ssresearch.2011.09.012

Sonnemann, J., & Goss, P. (2020). *COVID catch-up: Helping disadvantaged students close the equity gap* (Grattan Institute Report No. 2020-08, June 2020). Grattan Institute, Victoria. Retrieved April 14, 2021, from https://grattan.edu.au/wp-content/uploads/2020/06/COVID-Catch-up-Grattan-School-Education-Report.pdf

Sorin, R., & Iloste, R. (2006). Moving schools: Antecedents, impact on students and interventions. *Australian Journal of Education, 50*(3), 227–241. https://doi.org/10.1177/000494410605000302

Stevens, P. A., & Dworkin, A. G. (Eds.). (2019). *The Palgrave handbook of race and ethnic inequalities in education*. Palgrave Macmillan.

Stewart, H., Watson, N., & Campbell, M. (2018). The cost of school holidays for children from low income families. *Childhood, 25*(4), 516–529. https://doi.org/10.1177/0907568218779130

Stewart, S., te Riele, K., & Stratford, E. (2019). *Literacy teaching in Tasmania: Teaching practice and teacher learning*. Peter Underwood Centre, University of Tasmania. Retrieved June 7, 2021, from http://ecite.utas.edu.au/138381

Taylor, M., & Edwards, B. (2012). Housing and children's wellbeing and development: Evidence from a national longitudinal study. *Family Matters, 91*, 47–61.

te Riele, K. (2014). *Putting the jigsaw together: Flexible learning programs in Australia*. Victoria Institute for Education, Diversity and Lifelong Learning. Retrieved June 19, 2021, from https://vuir.vu.edu.au/31758/1/TheVicInstitute_PuttingTheJigsawTogether.pdf

Teo, Y. Y. (2018). *This is what inequality looks like*. Ethos Books.

Thompson, R. (2019). *Education, inequality and social class: Expansion and stratification in educational opportunity*. Routledge.

Tomaszewski, W., Zajac, T., Rudling, E., te Riele, K., McDaid, L., & Western, M. (2022). Uneven impacts of COVID-19 on the attendance rates of secondary school students from different socioeconomic backgrounds in Australia: A quasi-experimental analysis of administrative data. *Australian Journal of Social Issues, 00*, 1–20.

United Nations Educational, Scientific and Cultural Organization. (2021a). *Global monitoring of school closures*. Retrieved September 3, 2021, from https://en.unesco.org/covid19/educationresponse#schoolclosures

United Nations Educational, Scientific and Cultural Organization. (2021b). *Global duration of school closures*. Retrieved September 3, 2021, from https://en.unesco.org/covid19/educationresponse#durationschoolclosures

United Nations Educational, Scientific and Cultural Organization, United Nations Children's Fund, the World Bank, and Organisation for Economic Co-operation and Development. (2021). *What's next? Lessons on education recovery: Findings from a survey of ministries of education amid the covid-19 pandemic*. United Nations Educational, Scientific and Cultural Organization. Retrieved August 29, 2021, from https://en.unesco.org/gem-report/node/3512

van de Werfhorst, H. G. (2021). Inequality in learning is a major concern after school closures. *Proceedings of the National Academy of Sciences, 118*(20), 1–3. e2105243118.

Van Lancker, W., & Parolin, Z. (2020). COVID-19, school closures, and child poverty: A social crisis in the making. *The Lancet Public Health, 5*(5), e243–e244. https://doi.org/10.1016/S2468-2667(20)30084-0

Volante, L., Schnepf, S. V., Jerrim, J., & Klinger, D. A. (Eds.). (2019). *Socioeconomic inequality and student outcomes: Cross-national trends, policies, and practices* (Vol. 4). Springer Nature.

Warren, D., O'Connor, M., Smart, D., & Edwards, B. (2016). *A critical review of the early childhood literature*. Australian Institute of Family Studies. Retrieved July 2, 2021, from https://www.dese.gov.au/universal-access-early-childhood-education/resources/critical-review-early-childhood-literature

Winsper, C., Wolke, D., Bryson, A., Thompson, A., & Singh, S. P. (2016). School mobility during childhood predicts psychotic symptoms in late adolescence. *Journal of Child Psychology and Psychiatry, 57*(8), 957–966. https://doi.org/10.1111/jcpp.12572

CHAPTER 5

The Broader Role of Schools

Abstract Schools are increasingly expected to provide a holistic approach to education: one that addresses not only student learning, but also student wellbeing, social-emotional support and essential material support to children and young people. This comes with an expanding responsibility of schools to support diverse students' needs. Notable exemplars of models to redress disadvantage at school include the provision of breakfast clubs and interagency supports. As a result, school closures prompted by crisis, like those implemented throughout the COVID-19 pandemic can have ripple effects far beyond the delivery of formal education. Commonly, children and young people who require this broader school support tend to be most vulnerable to the effects of crisis.

This chapter examines the critical and under-recognised ways schools support vulnerable children and young people to be able to participate in learning, highlighting the potentially detrimental impacts of disaster-induced school closures to the wellbeing and achievement of this cohort. Therefore, during and after times of crises, there is a need for sustained responses to addressing educational inequities, such as through interagency collaboration or professional learning for educators as responders to student vulnerabilities as outlined in Chap. 2.

Keywords COVID-19 pandemic • School breakfast clubs • Material basics • Interagency collaboration • School systems • Vulnerability • Emotional wellbeing

E. S. Rudling et al., *Education and Equity in Times of Crisis*,
https://doi.org/10.1007/978-3-031-18671-4_5

Introduction

Schools provide many critical contributions to the development and wellbeing of children and young people; however, the full extent of these essential supports is not well known. In the last fifty years, we have seen educational systems respond to a variety of natural and human created crises. Natural disasters such as earthquakes, bushfires and hurricanes, pandemics such as COVID-19, have reinforced that it is "not until a crisis occurs such as a school closure, that people remember how intertwined local schools are with their lives" (Mutch, 2020, p. 4).

Despite its impacts, there is quite limited knowledge emerging over the wider role of schools and educational sectors in supporting vulnerable children and young people during times of crisis. Research after natural disasters like Hurricane Katrina (2005) in the United States (US) and the Christchurch Earthquakes (2010–11) in New Zealand reveal the role of schools during and after times of crises may be less about minimising the loss of student learning time, and more about the role schools play in emotional and social recovery (Australian Institute for Teaching and School Leadership, 2020). In relation to prolonged pandemics specifically, it is argued that there is "an unmet need for research that considers the multiplicity of potential impacts of school closures" (Kneale et al., 2020, p. 1).

This chapter foregrounds the expanding responsibility of schools in society and their role in providing a holistic approach to education—one that addresses not just student learning but also wellbeing, social-emotional support and essential material support to children and young people. It spotlights how school closures prompted by crisis have "ripple effects" far beyond the delivery of formal education (United Nations, 2020). This includes the reality that millions of children and young people rely on their schools for access to meals, psychosocial supports, social protection mechanisms and respite from other adverse circumstances at home (Baron et al., 2020; Brown et al., 2020).

In her seminal research into the role of schools in disaster response and recovery, Mutch (2020, p. 2) argues that it is critical that "schools are recognised for what they have undertaken over and above expectations, in crises contexts". True appreciation of all those involved in school settings, involves "a rethinking of social capital, resilience, individuals' roles and their contributions" (Nye, 2016, p. 88). Regarding the ways in which schools support children and young people, the importance of social capital cannot be discounted. Social capital refers to "networks together with shared norms, values and understandings that facilitate co-operation within or among groups" (Stiglitz et al., 2018, p. 284). It is the spoken and unspoken connections which facilitate cooperation, linkages, responses and innovations between people and networks. In school contexts, evidence of social capital is evident in the ways in which a school community works together to holistically support children and young people. During crises, this could not be more important.

School Meal Programs

Adequate nutrition is a cornerstone of childhood wellbeing. It is a confronting realisation that millions of children and young people worldwide experience food insecurity and rely on schools to provide basic nutrition and essential meals (Kelly & Hansel, 2020; United Nations, 2020). Schools are not only physical locations for the provision of food, but also act as critical conduits for organisations that deliver essential food and supplies to children and young people (see Foodbank, 2020; Jose et al., 2020; MacDonald, 2019).

Since the arrival of COVID-19, research has continued to evidence the significant role of schools in the provision of meals, as well as the challenges of school closures in ensuring children and young people still have access to essential food. For example, prior to COVID-19 in 2020 in the US, it is estimated that nearly 22 million children and young people received free or subsidised school meals (Kelly & Hansel, 2020). In Europe, more than six percent of households with children couldn't afford adequate nutritional meals in 2020 (Lancker & Parolin, 2020, p. 243). In Australia, there are limited figures on the number of school meals and nutrition programs that are provided nationally. However, research prior to COVID-19 lockdowns suggested that "one in five Australian parents (18 percent) live in households experiencing food insecurity" (Jose et al., 2020, p. 620) where they do not have regular access to sufficient, safe, and nutritious food to meet their dietary needs, and where children attend school at least one day a week without eating breakfast (Foodbank, 2020; Jose et al., 2020). Sanitarium's "Good Start Breakfast Club" program, a community-based partnership providing more than 2000 schools-based breakfast clubs across Australia, reported in the 2019/2020 financial year, that they had delivered 1,555,000 serves of food to students at more than 2000 schools (Sanitarium, 2020). In the Australian state of Victoria,

MacDonald's (2019) evaluation of the implementation of school-based breakfast programs in five hundred of the state's most disadvantaged primary schools showed 50,000 breakfasts are served every week.

The Benefits of School-Based Meal Programs

It is widely recognised that hungry children cannot learn effectively (Bayrakdar & Guveli, 2020; MacDonald, 2019). Children and young people in vulnerable households who are also experiencing food insecurity rely most on school meal programs. The provision of school meal programs and participation of children and young people in these programs are associated with a wide range of benefits (see Lancker & Parolin, 2020) that can "positively impact social, behavioural, and educational outcomes for children" (Graham et al., 2016, p. 12). The positive impacts of school meal programs on educational attainment have been witnessed in school breakfast clubs in Australia (see Byrne et al., 2018; Ichumar et al., 2018; Jose et al., 2020; MacDonald, 2019) and in the UK (Graham et al., 2016). Studies with children, parents and school staff identify similar themes across contexts which extend beyond the benefits of providing nutrition and sustenance alone to wider benefits, including opportunities for social support, safe spaces to foster connections with school staff and improved student capacity for learning and peer engagement (Bayrakdar & Guveli, 2020; Byrne et al., 2018; Ichumar et al., 2018).

MacDonald's (2019) research reveals that schools themselves identify strong links between the provision of breakfast clubs and their welfare and wellbeing practices. Other studies show that SBCs have the capacity to "develop social capital among students and provide unique opportunities to monitor student well-being" (Jose et al., 2020, p. 627). In our research (Brown et al., 2020), even outside of times of crises, the provision of meals and food to students is a much-needed support recognised by teachers, to support learning and wellbeing in children and young people: "…not having dinner, the evening before, nor breakfast in the morning. We provide breakfast programs, emergency lunches and fruit or vegetables (when available) for crunch 'n sip programs" (school staff cited in Brown et al., 2020, p. 38).

When students become separated from the support system that schools provide because of crisis, there is an obvious need for innovation and collaboration in how schools and communities can continue to provide students with access to essential food.

Responses and Innovations

Closures of schools during times of crisis disrupt and hamper the provision of essential services, such as meals and nutrition to children and young people, which exacerbates food insecurity and material deprivation of already vulnerable families and households (Lancker & Parolin, 2020). Depending on the duration of disruption, this can cause further marginalisation of those already vulnerable.

The COVID-19 pandemic, for example, has caused through the physical closures of schools, millions of children and young people to miss out on essential school-based meal programs. For example, the United Nations (2020) reports that in the first months of the COVID-19 pandemic, more than "370 million children in 195 countries were affected, increasing hunger and nutritional deficiencies for the most disadvantaged" (p. 10). During times of crisis, there is a pressing need to develop alternative ways of distributing essential food to children and young people who rely on school meals for basic nutrition and sustenance. United Nations (2020, p. 14) detail that many countries have recognised the importance of maintaining the provision of food for children and young people, including:

- More than 70 countries adapting their school meals programs to continue supporting children during school closures.
- Nearly 50 countries providing take-home rations to children and their families in various forms, including through daily meal delivery and pre-packaged monthly rations.
- 22 countries opting to replace school meals with vouchers or cash that families can use to buy food or other essential items.
- Reaching some 6.9 million learners in 45 low-income countries since the onset of the crisis with take-home rations by governments with the support of the UN system.

The United States has been particularly responsive to systemically tackling the needs of its 22 million children and young people who were recipients of school meals before COVID-19. New legislation fast tracked by State Governors has been introduced across states to establish "a nationwide waiver allowing schools to deliver school meals in different community settings" (Kelly & Hansel, 2020, p. 1). In some US states, this has included engaging NGOs to help distribute meals. This legislative intervention has been described by stakeholders as "a critical step toward ensuring that our students maintain access to the school meals they rely on for their health and well-being" (Kelly & Hansel, 2020, p. 1).

In Australia, our research showed that during school closures in 2020, concerns over children and young people not being able to access meal programs or essential supplies was a consistent theme. There were several examples of Australian schools providing food packages and essential resources to students known to be in need through deliveries to homes and pick up points (see Brown et al., 2020).

These responses and innovations provide a moment to reflect on the large-scale inequities in society that lead to so many children and young people being left to rely on schools for the most basic life needs such as essential meals. However, they also highlight that, after times of crises, we cannot overlook the fact that there will be additional students, who may not be on "an existing list" or known to schools because they are new to precarity, who find themselves hungry and vulnerable because of many households being crowded and under economic strain.

Not Safer at Home

Concern over child welfare during times of crises is prevalent in both public discourse and academic literature. There is ever increasing evidence that, during prolonged school closures, there are significant rises in the numbers of children and young people who will have witnessed or experienced abuse and violence (Cahill et al., 2020a, 2020b; Nicolson et al., 2020; Paul Ramsay Foundation, 2021). Increased rates of child abuse, neglect, and exploitation have also been reported during the Ebola outbreak in Africa from 2014 to 2016 (Lee, 2020), the 2008 Global Financial Crisis (Schneider et al., 2017) and the Australian bushfires (Parkinson & Zara, 2013).

Lockdown orders initiated to mitigate epidemiological impacts during COVID-19, cemented the reality that home is not necessarily the safest place for every child and young person. The irony of "stay home, stay safe" campaigns is clear when considering research showing that "home" is one of the most dangerous places for women and children worldwide (Humphreys et al., 2020; Kamenetz, 2020; Pfitzner et al., 2020; United Nations Office on Drugs and Crime [UNODC], 2020). The United Nations Population Fund (UNFPA) (2021) predicted that for every three months that COVID-19 related lockdowns continued, an additional 15 million cases of domestic violence will occur worldwide (also see Pfitzner et al., 2020; Paul Ramsay Foundation, 2021).

As discussed in Chap. 2, violence against women and child related abuse across the world has been referred to as the "shadow pandemic" (Dawuni, 2021; UNODC, 2020). Across countries, evidence is emerging that points to the impact of lockdowns and school closures including unprecedented rises in calls and contacts with domestic abuse hotlines and counselling supports in Australia (Boxhall et al., 2020; Nicolson et al., 2020), Kids Helpline (Batchelor et al., 2021; Nicolson et al., 2020) and increased demand for women's shelters across Australia (Paul Ramsay Foundation, 2021). Similar trends have been reported in the United Kingdom (see House of Commons Home Affairs Committee, 2020; Muller & Goldenberg, 2020; Pfitzner et al., 2020; Tuohy, 2020) and in Germany where there have been surges in demand for chat and telephone counselling for children and adolescents through specialist helplines (Jentsch & Schnock, 2020). In China, reports of domestic violence to police tripled during lockdown (Lee, 2020).

Schools as "Frontline" Child Welfare Responders

It is an uncomfortable proposition that a young person's school, not home, can be the safest place in their lives. Schools are seen to provide a protective environment to buffer less than ideal circumstances at home by acting as "essential protective resources" (Brown et al., 2020). Schools can be valuable sources of support and intervention for children and young people who have witnessed domestic violence or who are experiencing or at risk of family violence, abuse, and trauma (Davies & Berger, 2019).

Schools are vital to the early detection and identification of child welfare concerns. As well as actioning and providing support, we can indeed consider them to be first, frontline, and frequent responders to abuse, neglect, and trauma (Cahill et al., 2020b). Based on this proposition, we know that when students are not physically attending school, this hidden but essential function of schools is hampered. Prolonged crises limit the capacity and power of public agencies, including schools, to provide services and continuity of care to children and young people outside of the school environment. In some cases, children and young people may be "missing" for some time after schools reopen (Chrysanthos, 2020). Our research (Brown et al., 2020) indicates that school closures, especially over prolonged periods, involves removal of children and young people from "trusted adults" (e.g., teachers, counsellors, sports coaches) who are often observers and responders to child welfare and wellbeing concerns:

> Teachers and their daily interactions make up one of the key connections for students (particularly vulnerable students). They may only identify with one teacher and that teacher is a key point of connection for them. It is often their demeanour, appearance or absence from class that will trigger a teacher checking up on them… Where they identify or feel safe with a teacher, they will often seek them out, sometimes disclosing. Teachers also refer students who would otherwise be off the radar to the Social Worker. (Secondary school-based staff as cited in Brown et al., 2020, p. 35).

In many countries, school staff act as "mandatory reporters" and are legally required to report suspected child abuse and neglect to government authorities responsible for child welfare (Child Family Community Australia, 2020; St. John, 2013). Mandatory reporting is a term used to describe a legal requirement of particular or selected individuals or organisations, who, because of their professional role have an obligation to report suspected cases of child abuse and neglect to government authorities (St. John, 2013). The occupations most commonly acting as mandated reporters for children and young people are not surprisingly those who deal frequently with children and young people and include teachers, school social workers and psychologists, early childhood education and childcare practitioners as well as doctors, nurses, and police (Child Family Community Australia, 2020). There are significant variations in mandatory reporting requirements across countries, and within countries (see Child Family Community Australia, 2020). The most comprehensive mandatory reporting laws are in the USA, Canada, and Australia (St. John, 2013). In Australia, the Royal Commission into Institutional Responses to Child Sexual Abuse (2017) (re)highlighted the importance of mandatory reporting laws in the protection of children and young people.

Since schools collectively are front line mandatory reporters, school closures during times of crises effectively shut down these crucial reporting channels, restricting the ability of those responsible for child protection and wellbeing to report and monitor children and young people. For example, the loss of "established measures and strategies" (Jentsch & Schnock, 2020, p. 1) for protecting children's welfare has been raised since COVID-19 lockdowns began (see Baron et al., 2020). Prior to this, similar concerns existed in relation to school closures caused by natural disasters (Le Brocque et al., 2016; Lee, 2020; Parkinson & Zara, 2013). School staff (e.g., teachers, counsellors) as well as interagency staff involved in child welfare (e.g., children and youth services, police, social workers) emphasise the importance of *seeing* students in person: "Schools have eyes on kids. This is essential for identifying kids who may need child protection, and then for collaborating with other agencies to provide child safety support" (Advocacy Groups and Services officer, as cited in Brown et al., 2020, p. 35).

However, the opportunities to both thoroughly observe and engage with children and young people are drastically reduced during school closures. Interim measures (e.g., phone or video calls) do not always allow for issues to be detected or monitored: "often 'sighting' the child allows us to get some evidence or judgement, emerging cases of children who are very vulnerable but too scared to say over the phone or video what is happening" (Education system staff as cited in Brown et al., 2020, p. 35).

These concerns are consistent with Australian service providers' experiences of working with families and victims of domestic violence during COVID-19 lockdowns (Pfitzner et al., 2020). They report that victims become much more isolated because of a lack of face-to-face interactions, meaning "limited support available for families, no eyes on children and little opportunities to work with them" (Pfitzner et al., 2020, p. 15). Research examining the impact of the COVID-19 pandemic on children's access to support services and efforts to safeguard children in the UK also found that: "Even where the virtual element was judged to be working well, social workers were said to be concerned that they were missing things that would have been picked up in a face-to face visit" (Pearce & Miller, 2020, p. 288).

Responses and Innovations

As schools take on more responsibility for student wellbeing including monitoring child welfare, there is consensus that "schools alone" cannot always effectively help students without support and cross-agency collaboration (Steen & Noguera, 2010). There is no doubt that crises disrupt the provisioning of both school-based and interagency support for vulnerable and at-risk children and young people (Seymour et al., 2020).

Lessons from previous school closures due to crises highlight the critical importance of supporting integrated cross-sectoral approaches and partnerships with other agencies for the purposes of serving students experiencing multiple forms of disadvantage (Brymer et al., 2021). As frequent "first responders" in initiating interagency care, referrals, and monitoring for support and recovery (Cahill et al., 2020b; Le Brocque et al., 2016), there is a need to spotlight the ways in which schools and agencies can and do work together, to support children and young people. For example, findings and recommendations of the Victorian Royal Commission into Family Violence (2016) in Australia highlight the critical necessity of coordination across multiple agencies in order to share information and manage risks (Pfitzner et al., 2020, p. 18).

While there is limited published discussion of the implications of the COVID-19 pandemic for practitioners working to safeguard children (Pearce & Miller, 2020), across many contexts, cross-agency collaboration between schools and sectors has been prioritised, leaving them better placed to respond to crises in the future (Brown et al., 2020; Pearce & Miller, 2020). For example, since the COVID-19 pandemic, Canada has acknowledged the need for new child welfare policies which mandate states to develop and enact child protection strategies that encourage cross-sectorial coordination and involve children (Sistovaris et al., 2020, p. 30).

In the USA, child wellbeing taskforces, led by Departments of Education, have been set up in many states, charged with "supporting local leaders and communities in caring for students, particularly those who are vulnerable or most at risk" (National Governors Association, 2020, p. 1).

In the UK, schools, and agencies responsible for child welfare, have adapted and implemented new approaches to enabling continuity in supporting in the wellbeing of children and young people during crisis and disaster. There, the Association of Safeguarding Partners led a series of webinars for schools, police, and other relevant safeguarding agencies during COVID-19 related lockdowns in 2020. Over 80 organisations in the UK participated and shared information and practice innovations. They acknowledged the challenges caused by COVID-19 to their working environment, however there was "positive recognition" and consensus relating to the speed and intensity of communications, the development of working relationships between partners and the increased collaboration to solve problems relating to safeguarding children (Pearce & Miller, 2020, p. 288). Examples included more frequent "operational meetings" where "partner leads came together online to share concerns and explain how they were meeting them" (Pearce & Miller, 2020, p. 289). In one local partnership, lead partners (health, police, and schools) identified a list of vulnerable preschool children by sharing data from a range of service providers in areas of drug and alcohol support, domestic abuse, and primary care (Pearce & Miller, 2020; The Association of Safeguarding Partners, 2021).

In Australia, innovative responses to mitigating risks for known vulnerable children and young people were at the forefront of immediate action for schools. Our research indicates that educational sectors, schools, and interagency groups bringing together school sectors, police, criminal justice systems, children's services and NGOs met and worked more closely during COVID-19 lockdowns to ensure resources and support reached children and young people (Brown et al., 2020). As one participant whose education system role involved interagency liaison observed:

> We touch base every second day, if not every day, in one form or another. Those young people we are very worried about, we touch base every day and we try to get a visual either on Facetime or do a site visit to their home. We don't go into the house, but they come out and chat with us in the driveway. So, we try to eyeball them. (Education system staff as cited in Brown et al., 2020, p. 35)

We found a pervasive sense of hope that, despite the challenging circumstances, the crisis would enable those agencies to continue the high degree of collaboration and cooperation to ensure children's needs were being met both in the present and the future (Brown et al., 2020). What is most clear is a need to continue the "important cross jurisdictional and sectoral collaboration that has occurred in current educational policy and practice" (Brown et al., 2020, p. 55) during recovery from the COVID-19 pandemic.

While school resourcing alone cannot fix the underlying causes of wider societal inequities discussed in this book, there is a recognised need for resourcing collaboration and effective partnerships models between schools and various agencies to better address and respond to the needs of children and young people who are vulnerable and face multiple risk factors (Cooper et al., 2016; McMahon et al., 2000; Starkey, 2020). For example, coordinated and effective interagency collaboration should bring together key stakeholders such as schools, juvenile justice, housing, mental health, law enforcement and social work to address and care for the integrated needs of children and young people across multiple systems.

Serving the best needs of children with regards to child welfare must involve a consideration of their home, community, and school environments (Cooper et al., 2016) and be part of an ecosystem whereby schools work with wider institutions and structures which support society more broadly and vice versa. To this end, Bronfenbrenner's ecological systems theory (Bronfenbrenner, 1974, 1977, 1995) is useful in considering child development as a complex system of relationships affected by multiple levels of the surrounding environment, from immediate settings of family and school to broad cultural values, laws, and customs. As schools respond to more and more challenges and crises, the importance of interagency collaboration (Cooper et al., 2016) becomes ever more urgent.

"Schools as Anchors" for Building and Supporting Wellbeing of Children and Young People

Globally, the current picture of children and young people's poor mental health and wellbeing is complex and confronting. A significant proportion of children and adolescents are affected by mental health conditions, accounting for "16 percent of the global burden of disease and injury in people aged 10–19 years" (World Health Organisation [WHO], 2020). Record rates of depression, anxiety, self-harm and suicide in children and young people worldwide are widely reported (Ford et al., 2021; WHO, 2020). Evidence strongly suggests that the COVID-19 pandemic has greatly exacerbated these trends (Wong, 2021) with consistent reports of increased mental health concerns and conditions amongst children and young people because of lockdowns and wider uncertainty (see Levin, 2020; National Alliance on Mental Illness, 2021; Nicolson et al., 2020; Wong, 2021).

Questions are rightly raised over what can be done to best support and promote wellbeing amongst children and young people even outside of crisis, and this has often squarely fallen on educational systems and schools (Demkowicz & Humphrey, 2019; Muller & Goldenberg, 2020). According to the US Association for Children's Mental Health (2021), addressing mental health needs in school "is critically important because 1 in 5 children have a diagnosable emotional, behavioural or mental health disorder", while "1 in 10 young people have a mental health challenge that is severe enough to impair how they function at home, school or in the community" (Starkey, 2020, p.1; see further Kessler et al., 2005).

In recent years, the global wellbeing of children and young people has become "a core focus of social policy, with a growing imperative to locate well-being within the sphere of education" (Graham et al., 2016, p. 440). However, the efforts undertaken by schools to support wellbeing have not always been emphasised in educational policy initiatives (Humphrey, 2016; Kidger et al., 2015). These somewhat "hidden" aspects of schooling featured predominantly in our research; "with vulnerable students… the hidden or informal curriculum is needed as much as the formal one. It builds social skills, teamwork, confidence, resilience, and hope" (Advocacy Groups and Services officer, as cited in Brown et al., 2020, p. 33).

Effective and sustained school mental health initiatives can impact positively and significantly on students and school environments across several domains including learning, social relationships as well as wellbeing (see Demkowicz & Humphrey, 2019; Mutch, 2020; Mutch & Gawith, 2014). However, the resourcing and capacity of schools and their staff to provide these services is often strained.

Effects of Crises on the Provision of School-Based Wellbeing Services and Supports

School closures affect not only learning, but educational institutions at all levels and have contributed to the weakening of protective factors, including access to and delivery of essential psychosocial services and supports to vulnerable young people and children. This indeed prompts reflection, even outside of times of crisis, when considering the profound evidence that schools play a critical role in mental health and wellbeing as "platforms for prevention, diagnosis, and counselling" (United Nations, 2020, p. 10).

For children and young people with mental health needs a loss or disruption of support and services, through both short term and prolonged school closures, increases their vulnerability which can further hinder their inclusion in education (Batchelor et al., 2021; Organisation for Economic Co-operation and Development [OECD], 2020). There is increasing evidence which points to the detrimental impacts of crises and prolonged school closures on children and young people (Berger, 2020; Lee, 2020; Le Brocque et al., 2016; Muller & Goldenberg, 2020; Tomaszewski et al., 2022; WHO, 2020), noting that children are most vulnerable to "poor psychological outcomes following exposure to a range of potentially traumatic events" (Le Brocque et al., 2016, p. 1).

How and why physical school closures can destabilise children and young people and have an exacerbating effect on their mental health and wellbeing includes considerations of several factors related to the wider role of schools in their lives. For example, school routines, including the patterns and expectations of behaviour associated with attending and participating in school, are "important coping mechanisms for young people with mental health issues" (Lee, 2020, p. 421). When schools are physically closed to students, and these routines are uprooted, students "lose an anchor in life" (Lee, 2020, p. 421). Schools and teachers provide an important role in providing a sense of "normalcy and security" (Le Brocque et al., 2016, p. 1) and in monitoring and supporting student wellbeing and recovery. For already vulnerable and at-risk students experiencing trauma, stable routines, structures, and predictability provided by schools, are important foundations for learning: "…some have very difficult home lives and for them, school is a constant in their lives, it does not change, and they know what to expect each day" (Secondary school-based staff as cited in Brown et al., 2020, p. 36).

Loss of access to social supports, and school based mental health screening and wellbeing services, or exposure to stressful and traumatic circumstances at home (Muller & Goldenberg, 2020; Nicolson et al., 2020) can all lead to relapses in symptoms (Lee, 2020). Of course, crises and school closures not only have the potential to exacerbate already existing conditions, but they can also trigger and lead to new conditions amongst students who may not have previously had problems with mental health: Given the nature of the coronavirus pandemic, it's important to remember that these concerns could arise in relation to any child, not just those previously identified as vulnerable, and that many children will have had a range of challenging experiences (Cahill et al., 2020a, 2020b; National Alliance on Mental Illness, 2021; World Health Organisation, 2020).

While little is yet known about the longer-term effects of large-scale crises like the COVID-19 pandemic, on children and young people's mental health and wellbeing (Lee, 2020), the OECD (2021, p. 1) report that the "uncertainties and broad impacts" of COVID-19 have impacted young people to a greater degree, with young people in Belgium, France and the United States "30% to 80% more likely to report symptoms of depression or anxiety than adults".

An Australian educational and developmental psychologist found that students who have gone through lockdown from the COVID-19 pandemic "experience fear, anxiety, anger and sadness, reduced quality of sleep, restlessness and problems concentrating and hyperactivity and irritability" (Berger, 2020, p. 1). Cahill et al. (2020a, p. 10) cites that, "prolonged lockdowns can lead to psychological stress, diminished access to social support and increased rates of family violence" (see also Campbell, 2020). Similarly, Batchelor et al. (2021) and Nicolson et al. (2020) point to the clear detrimental impacts of COVID-19 pandemic related lockdowns on children and young people's mental health and wellbeing.

Schools provide more than learning, they provide protective factors including routine, stability, access to social support as well as extracurricular activities which contribute to wellbeing. When schools are closed, there are inevitable impacts on wellbeing:

> The shuttering of the American education system severed students from more than just classrooms, friends and extracurricular activities. It has also cut off an estimated 55 million children and teenagers from school staff members whose open doors and compassionate advice helped them build self-esteem, navigate the pressures of adolescence and cope with trauma. (Levin, 2020, para. 1)

School Readiness for Crisis and Recovery

Given the near-universal access that education provides to children and young people, it is posited that schools are the institution best suited to provide support to students and to mitigate the effects of trauma, abuse and grief experienced by children and young people (Ford et al., 2021; Muller & Goldenberg, 2020).

During times of crisis, the roles and expertise of all members of a school community can make significant contributions to immediate and ongoing responses and recovery (Mutch, 2020; Mutch et al., 2014; Nye, 2016). In addition to lead responders (such as school leaders and teaching staff), IT staff and allied health staff (such as school psychologists, social workers and counsellors, speech therapists, nurses, and chaplains) play an essential role in crisis response and recovery within school settings (Australian Institute for Teaching and School Leadership, 2020). School-based mental health professionals assume a particularly critical role in response and recovery (Cowan & Rossen, 2013) including collaborating with inter-agency providers and supports.

There are a range of issues relating to the role played by schools in effectively protecting and promoting the mental health and wellbeing of children and young people (see Demkowicz & Humphrey, 2019) during and after times of crisis. Once schools are fully reopened, they are charged with helping children and young people to recover and provide stability through helping children and young people to re-establish relationships and connections, and in supporting their wellbeing. However, this is not without challenges especially when larger groups of students are likely exposed to trauma, abuse, or violence (Cahill et al., 2020b; Paul Ramsay Foundation, 2021; United Nations, 2020) leaving teachers and school support staff presented with heightened collective needs amongst their students (Steen & Noguera, 2010).

Added to this, detrimental impacts on mental health and wellbeing stemming from crises aren't always directly evident after events. For some children and young people, distress and trauma may surface later: Cahill et al. (2020b, p. 4) noted "1 in 4 young people directly exposed to disasters still have post trauma symptoms three months after an event". Adjustment of most students is considered to take time and patience, however supporting students with mental health issues can involve additional workload including implementing trauma-informed practices. Our research found school staff needed to provide:

> Support for re-adjusting to the routines of school—they are likely to have less flexibility than they have become used to. … There could be various forms of stress associated with the circumstances they were in during the shutdown and additional counselling and social work support may be required. (Education system staff as cited in Brown et al., 2020, p. 37).

Teacher Training and Preparedness

While schools are advocated as key actors that support students following crises and trauma and noting that interventions delivered by school staff can be effective (Muller & Goldenberg, 2020), there is consensus that schools and staff need additional/specific training and support to effectively do this (Davies & Berger, 2019; Muller & Goldenberg, 2020). As frontline observers and responders, it is perhaps both naïve and unfair to assume that all school staff have the resources or training to deal with student wellbeing, even at the best of times, not the least in times of crisis.

Given their proximity to, and relationships with children and young people, teachers are seen as being in the most "unique and well-placed position to provide vital support" (Le Brocque et al., 2016, p. 1) to children and young people, and to model “appropriate social and emotional responses” (Cahill et al., 2020b, p. 9) during and following times of crisis.

Despite increases in trauma-informed practice in schools, it is unclear whether teachers are well-prepared to identify and support children with trauma backgrounds (Davies & Berger, 2019, p. 97). In her research into the role of schools in disaster response and recovery in New Zealand, Mutch (2020, p. 2) outlines one of her five key findings as “the multiple roles played by teachers from their first response to the ongoing support they provide for students and their families”. Unlike school psychologists, counsellors or nurses, the role of teachers in promoting and responding to emotional and wellbeing of their students is less specific, with teachers reporting a high degree of uncertainty about assisting students with trauma, as it is “simply beyond their role and expertise” (Le Brocque et al., 2016, p. 2). Recent research in Australia has shown that teachers feel “ill-equipped, unsupported and lack confidence” when it comes to managing student wellbeing and mental health during times of crisis (Berger, 2020, p. 1).

Teachers require access to timely and appropriate professional learning to enable them to support student wellbeing and fulfil their role as “first responders” (see Cahill et al., 2020b; Starkey, 2020). Muller and Goldenberg (2020, p. 22) report that formal professional support and training can empower teachers to not only better support students but to better cope themselves during times of crises. However, there is clearly an unmet need for professional training that better equips teachers to respond to trauma, vulnerability, and mental health issues in their students (Starkey, 2020). Additionally, there are important questions to be addressed about appropriate staffing levels in schools to enable such support to be provided. In particular, there is a growing need for the employment of school psychologists and counsellors on a continuing basis, who can support both students and staff, and develop the relationships needed to effectively address student needs. Not only do these allied professionals have the expertise needed, but enhanced integration of their work in schools also frees up teachers to focus on their own key role of providing learning support (Hunter & Sonnemann, 2022). Given the argument that “school crisis response is not a choice, it is inevitable” (Cowan & Rossen, 2013, para. 3), we are reminded of the critical role that schools play at these times and the need to support first responders in those environments.

Innovations and Responses

The scarce resources of many schools can make it more difficult to attend to the mental health and wellbeing needs of all young people and children, especially those with multiple risks factors (OECD, 2020). However there have been considerable innovations and adaptions in different countries implementing new platforms and forms to care to ensure continuity of services and support for children and young people. These have included expanded online and phone counselling and support services, e-mental health services by NGOs and online teletherapy and diagnostic assessments (see Colizzi et al., 2020; Goldschmidt, 2020; OECD, 2021; Racine et al., 2020).

There is also strong evidence that schools can provide psychoeducation supports to assist in both in the immediate aftermath of a crisis and as longer-term recovery efforts (Cahill et al., 2020b; Demkowicz & Humphrey, 2019; Ford et al., 2021). There have been calls for some time for mental health and emotional learning programs to be integrated parts of the curriculum or supported co-curricular activity (see Cahill et al., 2020b; Mutch, 2020; Mutch et al., 2014; Starkey, 2020). Social and emotional support programs also provide "children and young people with opportunities to safely process their experiences" (Mutch, 2020, p. 2). Social and Emotional Learning Programs (SELPs) can assist with mitigating the mental health impacts on students of crises in both short- and long-term interventions for all students, with targeted or specialised interventions for higher risk or vulnerable students (Cahill et al., 2020b).

Another critical point is that children and young people can be both highly vulnerable as well as highly resilient in crisis contexts (Cahill et al., 2020b) and should not be considered as "passive victims" (Mutch, 2020, p. 6). Across her research with schools following crises, Mutch (2020) argues all students have a right to participate in issues that relate to their experiences of crisis, and that "we need to listen to their views and engage them in suitable crisis recovery activities" (p. 6). Nicolson et al. (2020) also argue that children and young people need to be part of both the response to a crisis such as the COVID-19 pandemic as well as recovery efforts to promote resilience and build hope in a time of ongoing uncertainty (Nicolson et al., 2020). Berger (2020) points to the post-traumatic growth and resilience that can occur amongst children and young people following adversity, and that prompting students to share experiences and thoughts through well thought out interventions and programs can be beneficial.

School based interventions and opportunities can be powerful settings and effective environments to assist in building resilience and coping skills for everyday life (Ford et al., 2021). Likewise, partnerships with communities and regions (e.g., arts-based interventions) have also shown to be successful in developing programs that can assist schools in post crisis recovery (Cahill et al., 2020b; Mutch & Latai, 2019). For example, following both the New Zealand Christchurch Earthquakes (2010-11) and the Australian "Black Saturday" fires (2009) arts-based practices were employed to support schools and young people in affected regions (Cahill et al., 2020b). For example, illustrations made into books, large mosaics panels, documentaries and songs were examples of how collaborations with local community artists, schools, families, and young people were created to support wellbeing. Cahill et al. (2020b, p. 21) reports the positive impacts of such arts-based interventions on children and young people arguing that "the expressive power of the arts had supported their social connectedness and emotional processing of the experience" of crisis.

However, even outside of times of crisis, a major issue raised in the UK is "the additional lengths schools and colleges are having to go to in order to support vulnerable students—and that they are ill-equipped to do so" (Starkey, 2020). Thus, the resourcing and support of schools to provide mental health and wellbeing programs, the training of key school staff and the ability to create interagency partnerships is essential in considering how schools can continue to be effective in supporting the wellbeing of children and young people. In our concluding chapter we return to this issue.

Conclusion

As argued in earlier chapters, crises come in many forms but such events inevitably cause major disruptions to families, livelihoods, societies, and school communities (Cowan & Rossen, 2013). Most often than not, those children and young people who are vulnerable prior to a crisis, tend to be most vulnerable to its effects (Cahill et al., 2020b).

This chapter has foregrounded and celebrated some of the lesser known but critical roles of schools in building and protecting the wellbeing of children. Times of crisis present both opportunities and challenges. The COVID-19 pandemic has taught us that although we have "limited experience and evidence in responding to a situation of this magnitude" (Mental Health Foundation, 2021, p. 1), the need for providing essential resources, for safeguarding children, for delivering wellbeing interventions and services and the need for support for teachers and schools to take on these roles alone and in partnership is critical.

References

Australian Institute for Teaching and School Leadership. (2020). *Spotlight: The role of school leadership in challenging times*. Retrieved September 9, 2021, from https://www.aitsl.edu.au/research/spotlight/the-role-of-school-leadership-in-challenging-times

Baron, E. J., Goldstein, E. G., & Wallace, C. (2020). Suffering in silence: How COVID-19 school closures inhibit the reporting of child maltreatment. *Journal of Public Economics, 190*, 1–13. https://doi.org/10.1016/j.jpubeco.2020.104258

Batchelor, S., Stoyanov, S., Pirkis, J., & Kõlves, K. (2021). Use of Kids Helpline by children and young people in Australia during the COVID-19 pandemic. *Journal of Adolescent Health, 68*(6), 1067–1074. https://doi.org/10.1016/j.jadohealth.2021.03.015

Bayrakdar, S., & Guveli, A. (2020). *Inequalities in home learning and schools' provision of distance teaching during school closure of COVID-19 lockdown in the UK* (ISER Working Paper Series, No. 2020-09). Institute for Social and Economic Research (ISER), University of Essex. Retrieved September 10, 2021, from https://www.econstor.eu/bitstream/10419/227790/1/1703719352.pdf

Berger, E. (2020). Returning to school after lockdown: Six ways to support your students. *Monash University*. Retrieved December 1, 2020, from https://www.monash.edu/education/teachspace/articles/returning-to-school-after-lockdown-six-ways-to-support-your-students

Boxhall, H., Morgan, A., & Brown, R. (2020). *The prevalence of domestic violence among women during the COVID-19 pandemic: Statistical Bulletin no. 28*. Australian Institute of Criminology. https://www.aic.gov.au/publications/sb/sb28

Bronfenbrenner, U. (1974). Developmental research, public policy, and the ecology of childhood. *Child Development, 45*(1), 1–5. https://doi.org/10.2307/1127743

Bronfenbrenner, U. (1977). Toward an experimental ecology of human development. *The American Psychologist, 32*(7), 513–531. https://doi.org/10.1037/0003-066X.32.7.513

Bronfenbrenner, U. (1995). Developmental ecology through space and time: A future perspective. In P. Moen, G. H. Elder Jr., & K. Lüscher (Eds.), *Examining lives in context: Perspectives on the ecology of human development* (pp. 619–647). American Psychological Association. https://doi.org/10.1037/10176-018

Brown, N., te Riele, K., Shelley, B., & Woodroffe, J. (2020). *Learning at home during COVID-19: Effects on vulnerable young Australians* (Independent Rapid Response Report). Peter Underwood Centre for Educational Attainment, University of Tasmania. Retrieved May 26, 2020, from https://www.utas.edu.au/__data/assets/pdf_file/0008/1324268/Learning-at-home-during-COVID-19-updated.pdf

Brymer, M., Steinberg, A., Watson, P., & Pynoos, R. (2021). Prevention and early intervention programs for children and adolescents. In G. Beck & D. Sloan (Eds.), *The Oxford handbook of traumatic stress disorders* (pp. 1–533). Oxford University Press.

Byrne, M., Hill, S., Wenden, E., Devine, A., Miller, M., Quinlan, H., Shaw, T., Eastham, J., & Cross, D. (2018). *Evaluation of the Foodbank WA School Breakfast and Nutrition Education Program: Final Report.* Edith Cowan University, Telethon Kids Institute. Retrieved September 9, 2021, from https://www.foodbank.org.au/wp-content/uploads/2020/09/FBWA-2017-SBP-Evaluation-Report.pdf

Cahill, H., Dadvand, B., Shlezinger, K., Romei, K., & Farrelly, A. (2020a). *Natural disasters and pandemics - Supporting student and teacher wellbeing after crisis.* Melbourne Youth Research Centre, University of Melbourne.

Cahill, H., Shlezinger, K., Romel, K., & Dadvand, B. (2020b). *Research informed approaches to supporting student wellbeing post-disaster.* Youth Research Centre, Melbourne Graduate School of Education, The University of Melbourne. Retrieved August 8, 2021, from https://education.unimelb.edu.au/__data/assets/pdf_file/0006/3275115/YRC-Post-Disaster-Report.pdf

Campbell, A. (2020). An increasing risk of family violence during the Covid-19 pandemic: Strengthening community collaborations to save lives. *Forensic Science International: Reports, 2*, 100089. https://doi.org/10.1016/j.fsir.2020.100089

Child Family Community Australia. (2020). *Mandatory reporting of child abuse and neglect.* Australian Institute of Family Studies. Retrieved June 15, 2021, from https://aifs.gov.au/cfca/publications/mandatory-reporting-child-abuse-and-neglect

Chrysanthos, N. (2020, September 1). Thousands of NSW students never returned to school after lockdown. *Sydney Morning Herald.* Retrieved March 1, 2021, from https://www.smh.com.au/national/nsw/thousands-of-nsw-students-never-returned-to-school-after-lockdown-20200901-p55rai.html

Colizzi, M., Sironi, E., Antonini, F., Ciceri, M. L., Bovo, C., & Zoccante, L. (2020). Psychosocial and behavioral impact of COVID-19 in Autism Spectrum Disorder: An Online Parent Survey. *Brain Sciences, 10*(6), 341. https://doi.org/10.3390/brainsci10060341

Cooper, M., Evans, Y., & Pybis, J. (2016). Interagency collaboration in children and young people's mental health: A systematic review of outcomes, facilitating factors and Inhibiting factors. *Child: Care, Health and Development, 42*(3), 325–342. https://doi.org/10.1111/cch.12322

Cowan, K. C., & Rossen, E. (2013). Responding to the unthinkable: School crisis response and recovery. *Phi Delta Kappan, 95*(4), 8–12. https://doi.org/10.1177/003172171309500403

Davies, S., & Berger, E. (2019). Teachers' experiences in responding to students' exposure to domestic violence. *Australian Journal of Teacher Education, 44*(11), 96–109. https://doi.org/10.14221/ajte.2019v44.n11.6

Dawuni, J. (2021). *The gendered face of COVID-19: Women and access to justice.* United Nations Office of Drugs and Crime. Retrieved September 13, 2021, from https://www.unodc.org/dohadeclaration/en/news/2020/04/gendered-face-of-covid19-women-and-access-to-justice.html

Demkowicz, O., & Humphrey, N. (2019). *Whole school approaches to promoting mental health: What does the evidence say?* (Evidence Briefing No. 5). Manchester Institute of Education, University of Manchester. Retrieved August 19, 2020, from https://www.ucl.ac.uk/evidence-based-practice-unit/sites/evidence-based-practice-unit/files/evidencebriefing5_v1d7_completed_06.01.20.pdf.

Foodbank. (2020). *Foodbank Hunger Report 2020: Food insecurity in the time of COVID-19.* Retrieved March 1, 2021, from https://www.foodbank.org.au/wp-content/uploads/2020/10/FB-HR20.pdf?state=au

Ford, T., Degli Espoto, M., Crane, C., Taylor, L., Montero-Marin, J., Blakemore, S.-J., Bowes, L., Byford, S., Dalgeish, T., Greenberg, M., Nuthall, E., Phillips, A., Raja, A., Ukoumunne, O., Viner, R., Williams, M., Allwood, M., Aukland, L., Casey, T., & de Wilde, K. (2021). The role of schools in early adolescents' mental health: Findings from the MYRIAD Study. *Journal of the American Academy of Child & Adolescent Psychiatry.* https://doi.org/10.1016/j.jaac.2021.02.016

Goldschmidt, K. (2020). The COVID-19 pandemic: technology use to support the wellbeing of children. *Journal of Pediatric Nursing, 53*, 88–90. https://doi.org/10.1016/j.pedn.2020.04.013. https://www.ncbi.nlm.nih.gov/pmc/articles/PMC7161478/

Graham, A., Powell, M. A., Thomas, N., & Anderson, D. (2016). Reframing "well-being" in schools: The potential of recognition. *Cambridge Journal of Education, 47*(4), 439–455. https://doi.org/10.1080/0305764X.2016.1192104

House of Commons Home Affairs Committee. (2020). *Home Office preparedness for Covid-19 (Coronavirus): Domestic abuse and risks of harm within the home* (Session 2019–2021, Report No. 2). House of Commons, Parliament of the United Kingdom. Retrieved July 3, 2021, from https://publications.parliament.uk/pa/cm5801/cmselect/cmhaff/321/321.pdf

Humphrey, N. (2016). The role of schools in promoting children's mental health. In D. Morley (Ed.), *Critical Issues in Child and Adolescent Mental Health* (pp. 129–147). Palgrave.

Humphreys, K. L., Myint, M. T., & Zeanah, C. H. (2020). Increased risk for family violence during the COVID-19 pandemic. *Pediatrics, 146*(1). https://doi.org/10.1542/peds.2020-0982

Hunter, J., & Sonnemann, J. (2022). *Making time for great teaching: How better government policy can help.* Grattan Institute.

Ichumar, S., Dahlberg, E., Paynter, E., Lucey, F., Chester, M., Papertalk, L., & Thompson, S. (2018). Looking through the keyhole: Exploring realities and possibilities for school breakfast programs in rural Western Australia. *Nutrients, 10*(3), 371. https://doi.org/10.3390/nu10030371

Jentsch, B., & Schnock, B. (2020). Child welfare in the midst of the coronavirus pandemic—Emerging evidence from Germany. *Child Abuse & Neglect, 110*(2), 104716. https://doi.org/10.1016/j.chiabu.2020.104716

Jose, K., MacDonald, F., Vandenberg, M., Williams, J., Abbott-Chapman, J., Venn, A., & Smith, K. J. (2020). School breakfast club programs in Australian primary schools, not just addressing food insecurity: A qualitative study. *Health Education & Behavior, 47*(4), 619–630. https://doi.org/10.1177/1090198120920193

Kamenetz, A. (2020). *Child sexual abuse reports are on the rise amid lockdown orders.* NPR. Retrieved May 22, 2020, from https://www.npr.org/sections/coronavirus-live-updates/2020/04/28/847251985/child-sexual-abuse-reports-are-on-the-rise-amid-lockdown-orders

Kelly, J., & Hansel, K. (2020, November 3). Coronavirus: What child welfare systems need to think about. *The Imprint.* Retrieved September 19, 2021, from https://imprintnews.org/child-welfare-2/coronavirus-what-child-welfare-systems-need-to-think-about/41220

Kessler, R. C., Berglund, P., Demler, O., Jin, R., Merikangas, K. R., & Walters, E. E. (2005). Lifetime prevalence and age-of-onset distributions of DSM-IV disorders in the National Comorbidity Survey Replication. *Archives of General Psychiatry, 62*(6), 593–510. https://doi.org/10.1001/archpsyc.62.6.593

Kidger, J., Heron, J., Leon, D. A., Tilling, K., Lewis, G., & Gunnell, D. (2015). Self-reported school experience as a predictor of self-harm during adolescence: A prospective cohort study in the South West of England (ALSPAC). *Journal of Affective Disorders, 173,* 163–169. https://doi.org/10.1016/j.jad.2014.11.003

Kneale, D., O'Mara-Eves, A., Rees, R., & Thomas, J. (2020). School closure in response to epidemic outbreaks: Systems-based logic model of downstream impacts. *F1000Research, 9*(1), 352. https://doi.org/10.12688/f1000research.23631.1

Lancker, W. V., & Parolin, Z. (2020). COVID-19, school closures, and child poverty: A social crisis in the making. *The Lancet Public Health, 5*(5), e243–e244. https://doi.org/10.1016/S2468-2667(20)30084-0

Le Brocque, R., De Young, A., Montague, G., Pocock, S., March, S., Triggell, N., Rabaa, C., & Kenardy, J. (2016). Schools and natural disaster recovery: The unique and vital role that teachers and education professionals play in ensuring the mental health of students following natural disasters. *Journal of Psychologists and Counsellors in Schools, 27*(1), 1–23. https://doi.org/10.1017/jgc.2016.17

Lee, J. (2020). Mental health effects of school closures during COVID-19. *The Lancet Child & Adolescent Health, 4*(6), 421. https://doi.org/10.1016/S2352-4642(20)30109-7

Levin, D. (2020, May 20). In a world "so upside down," the virus is taking a toll on young people's mental health. *The New York Times.* Retrieved September 15, 2020, from https://www.nytimes.com/2020/05/20/us/coronavirus-young-people-emotional-toll.html

MacDonald, F. (2019). *Evaluation of the school breakfast clubs program final report.* The Institute for Sustainable Industries and Liveable Cities, VU Research, Victoria University. https://www.vu.edu.au/sites/default/files/evaluation-school-breakfast-clubs.pdf.

McMahon, T., Ward, N., Kline Pruett, M., Davidson, L., & Griffith, E. (2000). Building full-service schools: Lessons learned in the development of inter-agency collaboratives. *Journal of Educational and Psychological Consultation, 11*(1), 65–92. https://doi.org/10.1207/s1532768Xjepc1101_05

Mental Health Foundation. (2021). *The challenge facing schools and pupils.* Retrieved November 9, 2020, from https://www.mentalhealth.org.uk/coronavirus/returning-school-after-lockdown/challenge-facing-schools-pupils

Muller, L. -M., & Goldenberg, G. (2020). *Education in times of crisis: The potential implications of school closures for teachers and students.* Chartered College of Teaching. Retrieved February 19, 2021, from https://chartered.college/education-in-times-of-crisis-the-potential-implications-of-school-closures-for-teachers-and-students/

Mutch, C. (2020). How might research on schools' responses to earlier crises help us in the COVID-19 recovery process? *Set: Research Information for Teachers, 2*, 3–10. Retrieved June 27, 2020, from https://www.nzcer.org.nz/system/files/journals/set/downloads/Set%202020_2_003_0.pdf

Mutch, C., & Gawith, E. (2014). The New Zealand earthquakes and the role of schools in engaging children in emotional processing of disaster experiences. *Pastoral Care in Education, 32*(1), 54–67. https://doi.org/10.1080/02643944.2013.857363

Mutch, C., Gibbs, L., O'Connor, P., & MacDougall, C. (2014). Editorial. *Pastoral Care in Education, 32*(1), 1–4. https://doi.org/10.1080/02643944.2014.881599

Mutch, C., & Latai, L. (2019). Creativity beyond the formal curriculum: Arts-based interventions in post-disaster trauma settings. *Pastoral Care in Education, 37*(3), 230–256. https://doi.org/10.1080/02643944.2019.1642948

National Alliance on Mental Illness. (2021). *School during the pandemic: Mental health impacts on students*. National Alliance on Mental Illness, California. Retrieved September 19, 2021, from https://namica.org/blog/impact-on-the-mental-health-of-students-during-covid-19/

National Governors Association. (2020). *Memorandum: Addressing the decline in child abuse reports and supporting child well-being*. Retrieved July 15, 2020, from https://www.nga.org/center/publications/child-abuse-reporting-covid-19/

Nicolson, S., Newell, S., & Collyer, B. (2020). *Impacts of COVID-19 on children and young people who contact Kids Helpline (2020)*. Australian Human Rights Commission and yourtown. Retrieved January 5, 2021, from https://humanrights.gov.au/our-work/childrens-rights/publications/impacts-covid-19-children-and-young-people-who-contact-kids

Nye, A. (2016). Working from the boot of a red Falcon: The impact of major fires in four Australian schools. *Australian and International Journal of Rural Education, 26*(1), 83–98. Retrieved December 3, 2020, from https://eric.ed.gov/?id=EJ1240749

Organisation for Economic Co-operation and Development. (2020). *The impact of COVID-19 on student equity and inclusion: Supporting vulnerable students during school closures and school re-openings*. Retrieved January 12, 2021, from https://www.oecd.org/coronavirus/policy-responses/the-impact-of-covid-19-on-student-equity-and-inclusion-supporting-vulnerable-students-during-school-closures-and-school-re-openings-d593b5c8/

Organisation for Economic Co-operation and Development. (2021). *Supporting young people's mental health through the COVID-19 crisis*. Retrieved May 21, 2021, from https://www.oecd.org/coronavirus/policy-responses/supporting-young-people-s-mental-health-through-the-covid-19-crisis-84e143e5/

Parkinson, D., & Zara, C. (2013). The hidden disaster: Domestic violence in the aftermath of natural disaster. *Australian Journal of Emergency Management, 28*(2), 28–35. Retrieved February 22, 2021, from https://www.researchgate.net/publication/288432753_The_hidden_disaster_Domestic_violence_in_the_aftermath_of_natural_disaster

Paul Ramsay Foundation. (2021). *Shelter from the Storm: Learning from the rise in domestic and family violence during COVID-19*. Retrieved September 20, 2021, from https://paulramsayfoundation.org.au/wp-content/uploads/2021/06/PRF-Shelter-Report-2021-V12-June.pdf

Pearce, J., & Miller, C. (2020). Safeguarding children under Covid-19: What are we learning? *Journal of Children's Services, 15*(4), 287–293. https://doi.org/10.1108/JCS-06-2020-0021

Pfitzner, N., Fitz-Gibbon, K., & True, J. (2020). *Responding to the "shadow pandemic": Practitioner views on the nature of and responses to violence against women in Victoria, Australia during the COVID-19 restrictions*. Monash

Gender and Family Violence Prevention Centre, Monash University. Retrieved January 3, 2021, from https://apo.org.au/sites/default/files/resource-files/2020-06/apo-nid306064.pdf

Racine, N., Hartwick, C., Collin Vézina, D., & Madigan, S. (2020). Telemental health for child trauma treatment during and post-COVID-19: Limitations and considerations. *Child Abuse & Neglect, 110*(Part 2), 1–9. Retrieved June 1, 2022, from https://doi.org/10.1016/j.chiabu.2020.104698

Sanitarium. (2020, June). *Good Start Breakfast Clubs.* https://www.sanitarium.com.au/social-purpose/tackling-hunger/good-start-breakfast-club/good-start-breakfast-clubs

Schneider, W., Waldfogel, J., & Brooks-Gunn, J. (2017). The Great Recession and risk for child abuse and neglect. *Children and Youth Services Review, 72*, 71–81. https://doi.org/10.1016/j.childyouth.2016.10.016

Seymour, K., Skattebol, J., & Pook, B. (2020). Compounding education disengagement: COVID-19 lockdown, the digital divide and wrap-around services. *Journal of Children's Services, 15*(4), 243–251. https://doi.org/10.1108/JCS-08-2020-0049

Sistovaris, M., Fallon, B., Miller, S., Birken, C., Denburg, A., Jenkins, J., Levine, J., Mishna, F., Sokolowski, M., & Stewart, S. (2020). *Child welfare and pandemics: Literature scan.* Policy Bench, Fraser Mustard Institute for Human Development, University of Toronto. Retrieved March 10, 2021, from https://alliancecpha.org/en/system/tdf/library/attachments/child-welfare-and-pandemics-literature-scan_optimized.pdf?file=1&type=node&id=37657

St. John, E. (2013). *Mandatory reporting of child abuse around the world.* AO Advocates. Retrieved September 13, 2021, from https://aoadvocates.com/blog/2013/12/5/mandatory-reporting-of-child-abuse-around-the-world

Starkey, H. (2020). *How we can support mental health in schools.* True Education Partnerships. Retrieved March 9, 2021, from https://www.trueeducationpartnerships.com/schools/tips-for-supporting-mental-health-in-schools/

Steen, S., & Noguera, P. A. (2010). A broader and bolder approach to school reform: Expanded partnership roles for school counselors. *Professional School Counseling, 14*(1), 42–52. https://doi.org/10.1177/2156759X1001400105

Stiglitz, J., Fitoussi, J., & Durand, M. (Eds.). (2018). *For good measure: Advancing research on well-being metrics beyond GDP.* OECD Publishing. https://doi.org/10.1787/9789264307278-en

The Association of Safeguarding Partners. (2021). *The Association of Safeguarding Partners.* Retrieved September 13, 2021, from https://www.theASP.org.uk

Tomaszewski, W., Zajac, T., Rudling, E., te Riele, K., McDaid, L., & Western, M. (2022). Uneven impacts of COVID-19 on the attendance rates of secondary school students from different socioeconomic backgrounds in Australia: A quasi-experimental analysis of administrative data. *Australian Journal of Social Issues, 00*, 1–20.

Tuohy, W. (2020, April 12). Helpline calls by family violence perpetrators "sky-rocket" amid isolation. *The Age*. Retrieved April 13, 2020, from https://www.theage.com.au/national/victoria/helpline-calls-by-family-violence-perpetrators-skyrocket-amid-isolation-20200410-p54iw7.html

United Nations. (2020). *Policy brief: Education during COVID-19 and beyond*. Retrieved October 4, 2020, from https://www.un.org/development/desa/dspd/wp-content/uploads/sites/22/2020/08/sg_policy_brief_covid-19_and_education_august_2020.pdf

United Nations Office on Drugs and Crime. (2020). *The shadow pandemic: COVID-19 and essential services for women and girl survivors of violence*. Retrieved September 13, 2021, from https://www.unodc.org/documents/Advocacy-Section/EssentialServices-Infographic-COVID-FIN_WEB_VERSION.pdf

United Nations Population Fund. (2021). *Update on UNFPA response to the COVID-19 pandemic*. Retrieved May 2, 2021, from https://www.unfpa.org/sites/default/files/board-documents/main-document/FINAL_UNFPA_response_to_COVID-19_pandemic_-_vffs.pdf

Wong, R. (2021, August 4). As COVID-19 rose, so did teen suicide attempts. Girls are most at risk. *Houston Chronicle*. Retrieved September 10, 2021, from https://www.houstonchronicle.com/lifestyle/renew-houston/health/article/As-COVID-19-rose-so-did-teen-suicide-attempts-16362650.php

World Health Organization. (2020). *Adolescent mental health*. Retrieved January 19, 2021, from https://www.who.int/news-room/fact-sheets/detail/adolescent-mental-health

CHAPTER 6

New Learning Provisions and Pedagogies

Abstract The global COVID-19 pandemic shifted the locus of education from the school to a shared responsibility between school and home. This disaggregation of learning highlights both the holistic approach offered by schools in school-based learning (as outlined in Chap. 4), and the adaptation required by students and families to support learning from home. This chapter examines the challenges and opportunities of delivering an educational program from school at home, and what we have learnt. The reliance on technological delivery of learning during the COVID-19 pandemic has cleaved wider the digital divide where a lack of access, and unfamiliarity with learning platforms, undermine equity in blended learning models. However, if done well, technology enhanced learning in conjunction with blended learning has the potential to engage, support, and extend learners to make access to education more equitable both inside and outside times of crisis.

Keywords COVID-19 pandemic • Learning from home • Digital education • Digital divide • Blended learning • Technology • Technology enhanced learning • Digital inclusion • Digital equity

E. S. Rudling et al., *Education and Equity in Times of Crisis*,
https://doi.org/10.1007/978-3-031-18671-4_6

Introduction

The COVID-19 pandemic saw almost 190 countries close their schools for instruction and move to support learning at home during periods of lockdown. This has affected around 1.5 billion students worldwide (Organisation for Economic Co-operation and Development [OECD], 2021). The scale of this shift from school to home education is unprecedented and has sharpened the global focus on how education is facilitated and delivered to children and young people.

During COVID-19, the move to learning at home required a swift response from teachers to plan for delivery of curriculum off-site, particularly (but not exclusively) through the medium of technology. This transition was supported in a range of ways, inclusive of online platforms (via a device), use of mobile phones, television, radio, or more traditional take home schoolwork packages (OECD, 2021).

A world seemingly awash with technology, and a plethora of online learning materials, many of which are open access, may superficially appear to provide a "ready-made" platform to move learning from a classroom or school setting. The idea of learning through "any content, anytime, anywhere, any device" (Altmann et al., 2014) is a catchcry of tech enthusiasts and businesses, but the experiences of the COVID-19 pandemic affirmed that this capability was insufficient to enable a rapid and effective translation of learning away from classrooms, especially for students from disadvantaged backgrounds.

Nevertheless, the possibilities opened by technology, not only in terms of innovation in content delivery but also in its capacity for facilitating individualised and inclusive learning, provide a solid foundation for thinking differently about future facilitation of education. The promise of such innovation has been recognised by the United Nations (2020) and has catalysed educators and educational researchers to begin to contemplate how this can be capitalised upon. This includes consideration of what is needed to provide a foundation for successful learning, what curriculum is taught; how learning is facilitated; and where learning can take place (Dabrowski et al., 2020; Zhao & Watterson, 2021).

Disaggregation of the Learning Environment

Any discussion of what has been learnt due to the extensive shift from school to home needs to consider how the learning environment changed during this shift. Although many would subscribe to the notion that learning happens everywhere, the school as a site of learning is both well accepted by the public and supported through government policy settings and funding arrangements. Support for public education broadly, and for schools specifically, recognises the importance placed by governments on universal access to education (Harrington, 2011). The learning environment, as broadly conceived and centred through the schooling system, has several inter-related and interdependent domains that we argue are necessary to support success in learning for children and young people (see Fig. 6.1).

Whilst these domains could be viewed as always being a shared responsibility between home and school, learning at home has not only shone a spotlight on individual constituent elements within the domains, but also on the balance of provision. In usual times, the supports listed in italics are

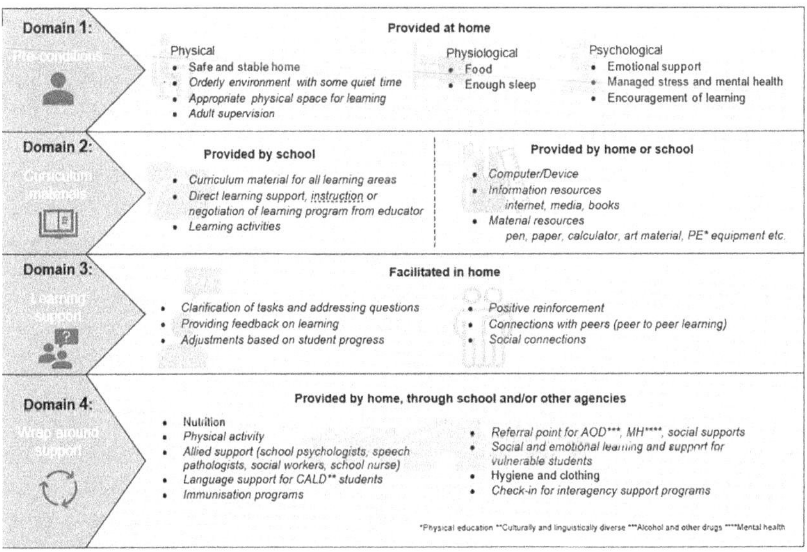

Fig. 6.1 Domains that support learning across different contexts. Figure adapted from Brown et al. (2020, p. 27)

predominantly provided by schools. The need for these to be provided by, or facilitated in, the home environment demonstrates the extent of the disaggregation and shift of responsibility to parents and caregivers (Brown et al., 2020; Fay et al., 2020).

It is important to recognise, however, that this shift does not represent an abrogation of responsibility by schools towards parents/carers. The data from 2020–21 strongly supports a rapid, responsive, and nuanced reaction from schools and school systems to provide the best possible support to parents and caregivers with these new responsibilities.

Online Provision

In the early phases of learning at home, there was a focus on the provision of curriculum materials and learning activities for students. In many cases this was delivered through a learning management system (LMS) that was already in place in a school or system.

Whilst moving curriculum content online could rapidly take advantage of the plethora of high-quality digital resources that are available, a key determinant is access. This access is twofold—both to appropriate devices and to reliable, affordable internet. During the COVID-19 crisis this was brought into sharp relief across the globe (e.g., Australia [Australian Broadband Advisory Council, 2020]; Pakistan [Adnan & Anwar, 2020]; Georgia [Basilaia & Kvavadze, 2020]; Singapore [Lee & Yeo, 2020]; and US [Fay et al., 2020]). Unsurprisingly, issues of access were most evident in families who are already experiencing some form of disadvantage (Bubb & Jones, 2020) as one school-based staff member in our research explained:

> Many students have been loaned laptops and there are challenges regarding internet access—no internet connection, geographically isolated location—poor coverage.

In Australia, only 68% of school-aged children living in disadvantaged communities have internet access at home, compared to 91% of students living in advantaged communities (Graham & Sahlberg, 2020). Even within these communities, some cohorts of students were even further likely to be disadvantaged according to an advocate who works with people with disabilities:

> Poverty and disability go hand in hand, having internet or using internet for learning may actually very well take away from other resources they need. Technological literacy is taken for granted completely in current discussions, both through caregivers as well as young people with disability. (Advocacy Groups and Services officer)

Even in households where there was internet, the findings of the Australian Digital Inclusion Index reveal that this most often stemmed from a mobile phone (Thomas et al., 2020). Coupled with the limitation of a phone as a learning device was the relative expense of mobile data—again disproportionately affecting disadvantaged, Indigenous and remote households (Thomas et al., 2020). This was also confirmed in our research where a primary school-based staff member highlighted the issues of disproportionate access to online learning:

> Although the majority of families do have an internet connection, this is predominantly through the use of their phones as the connection to the internet. Very few have the NBN[1] or another internet provider. The data is burnt through quickly if there is more than one child who is connecting a device. As a large majority of phones are PAYG[2], this is a significant financial burden.

The response from many schools, school systems, governments and not-for-profit groups was to provide devices, and mobile internet to families (Brown et al., 2020). However, it was not only access to a device and internet that provided a challenge to widespread curriculum delivery via the internet. Online delivery was found to be more effective when students (and their caregivers) had a degree of familiarity and confidence with the platform. In some cases, this was built post hoc through tailoring curriculum content to the whole family as an education system staff member explained:

> …we worked behind the scenes with schools to provide them with a plethora of online resources that could be used by families, so much work with families, beautiful webinars and sessions specifically for families and helping them to work with the school.

[1] National Broadband Network (Australia).

[2] Pay As You Go.

Offline Provision

Where students did not have access to a device, or to the internet, other media were used. This included use of USB memory sticks for stand-alone devices; and paper copy activities or learning packs that supported student learning activities. School staff members reported that, despite the urgency in the situation, there was some ability to differentiate material and tailor activities to specific student need.

> Overwhelmingly the families want hard copy materials because they do not have reliable data or consistent access to laptops for all household members. We created a data wall of the student faces showing what type of learning provision is suited to them. Very few want anything internet-based due to the cost and lack of reliability of access to this along with limited device access. (Secondary school-based staff as cited in Brown et al., 2020, p. 43).

For many schools, the need for additional material support for learning activities was also clear from the very early stages of learning at home, even when activities were not internet-based:

> We put [the packs] together so when we sent activities, we knew that they had what they needed, [for example] dice, chalk to go out play outside and draw on the cement. Because it is 2 hours away for parents to go to a stationery shop or to go to Kmart [department store] to find materials we provided everything. It was an investment that we thought was worth it for our children. (School-based leader of an isolated school as cited in Brown et al., 2020, p. 44)

Acknowledging the access challenges associated with using computers, tablets or even mobile phones for learning, there was also widespread use of more traditional technologies to supplement, or to substitute for, access to other media. In their Rapid Response report, Reimers and Schleicher (2020) surveyed 330 respondents across 98 countries and found a significant number of countries used television or radio to supplement learning at home.

Television and Radio

Across the world, UNESCO identified the use of television to support home learning in over 80 countries (OECD, 2021). These included Argentina, Austria, Belgium, Costa Rica, Croatia, Czech Republic,

Finland, Georgia, Indonesia, Israel, Latvia, Malaysia, Romania, Russia, Saudi Arabia, and the United States, where local children's TV producers and distributors contributed to home-based learning (ReliefWeb, 2020). Turkey established three television channels to deliver educational content for each key phase of schooling (Öçal et al., 2021) and the South African department of education delivered live lessons to learners through television, desktop computer, as well as radio (Mhlanga & Moloi, 2020).

In Australia, the public broadcaster the Australian Broadcasting Commission (ABC) is a long-standing contributor to the education of Australian children and pedagogy of local teachers, caregivers, and parents (Balanzategui et al., 2021). ABC outlets responded to school closures with the provision of formal education materials to supplement the broadcaster's existing suite of services and further support the delivery of school curricula. Departments of Education in Australian states also made explicit reference to selected ABC Kids[3] materials (Giuffre, 2021).

Radio was also used extensively worldwide (OECD, 2021). This medium has the advantage of reaching students in areas where internet is inaccessible, patchy, or too expensive. For example, in the remote Australian town of Wilcannia, where broadband was not available and mobile internet both unreliable and unaffordable, daily radio broadcasts where teachers read stories and explained daily tasks supplemented the hard copy activities delivered to homes (Cousins, 2020). A program in Argentina was developed to deliver 14 hours of television and seven hours a day of radio content each day to supplement hard copy learning resources (ReliefWeb, 2020).

Content Provision Is Not Learning

However, content provision alone is not sufficient for learning, it is what the student does with that content that is important. This brings domain 3 of Fig. 6.1 (see above) into focus. In essence this domain is pedagogy, the strategies that are used by teachers to actively engage students in the learning process. This includes instruction, questioning, differentiation of material, feedback, and facilitating peer to peer interactions. With the rapid shift to learning at home, the practices of the classroom needed to be translated and adapted to a new site of learning, with a recognition that these may be asynchronous. Before COVID-19, classroom life largely

[3] Part-time channel airing at specific times of the day showing children's programs.

revolved around synchronous instruction—that is to say that a classroom of students learning similar content at the same time. Asynchronous learning is a phenomenon that has taken on new significance during the COVID-19 pandemic, with lessons pre-recorded or delivered to smaller groups or individual students at different times. A consideration of the new, and different interactions in this environment is also necessary.

Instructional design can inform online learning provision and incorporate both synchronous and asynchronous interactions between students their teachers and their peers (Brown et al., 2013). However, designing curriculum to minimise the necessity of parental/carer involvement cannot preclude the need to establish a learning environment, motivate learners, troubleshoot technology, and provide timely feedback for many learners at home. Moreover, depending upon the number and age of children in households, the capacities of caregivers to provide this support has been highly variable. Coyne et al. (2020) argue that although structure and predictability may be helpful, suddenly implementing a high level of structure in a context that is not normally so structured (e.g., home), is likely to be experienced as adding to the stress of the situation and will be resisted and avoided by many. Similar themes were identified in our research amongst school-based staff:

> Students may feel overwhelmed by this new way of learning, and some will find it difficult to self-motivate and plan and organise their time. I think there will be a great deal of conflict between parents and their children during this time, particularly if parents are trying to work from home and support their child's learning. (Secondary school-based staff as cited in Brown et al., 2020 p. 22).

> The biggest problem that parents are reporting is managing behaviour. Getting kids to settle and do work is very difficult. Some parents have been requesting support on how to help kids get on task. Parents report spending a lot of time on managing behaviour and disruptions in the home environment. A number of parents have very low levels of literacy themselves (40–50%), and this is compounding the challenge. (Primary school-based staff as cited in Brown et al., 2020, p. 22)

Critically, although initially focussed on the provision of curriculum materials, and learning activity, support from educational institutions also extended into the other two domains of pre-conditions (domain 1) and wrap around support (domain 4), with a particular focus on vulnerable students. Some schools utilised text messaging, video, or phone calls to stay connected with families, and for many this was systematised early in the pandemic.

> In the case of Pastoral Care priority students [rural, special learning needs, students with health issues or low socioeconomic status (SES) students], the House Head or Counsellor will make individual contact based on a priority list. (School-based staff cited in Brown et al., 2020, p. 44)

In domain 1 of Fig. 6.1, there have been some aspects of the learning environment that provided clear challenges, and which were not easily overcome. Finding a quiet space to study or learn at home is difficult for many families (also see Chap. 5). This disproportionally affects students who were already experiencing challenges to learning with overcrowding recognised as being more prevalent in households in lower SES and remote and very remote areas (Australian Institute of Health and Welfare [AIHW], 2021). Nevertheless, where it was within the sphere of influence, schools have responded quickly, providing food packs to families, and connecting families to support services where possible (Cullinane & Montacute, 2020; Eurochild, 2020; UNESCO, 2020).

The elements of wrap around support in domain 4 are arguably those that differentially affect vulnerable students. Students with disability, learning challenges, or who are connected to allied services through school were unquestionably disadvantaged through periods away from school as discussed in Chaps. 2 and 5 (see Dickinson & Yates, 2020; McDaid et al., 2020).

Key Developments

A consideration of the challenges and opportunities reported through the COVID-19 pandemic has illuminated key learnings that have implications not only for any future crises but also to inform our current practices to better support a diversity of learners. The identification of these challenges, and the possibilities for addressing them, has the potential to increase equity in education (Sahlberg, 2021). Each of these challenges and opportunities is now outlined and discussed.

1. *The supports for learning offered by a school go beyond curriculum and pedagogy*

While consideration of curriculum content provision, and support for learning was an early focus, the diverse needs of individual students quickly emerged during learning at home. The programs offered by schools that supplemented and complemented the learning program, school breakfasts and lunches; health and wellbeing activities and connections to allied professionals and agencies were no longer operational. This created heightened concern amongst educators which we described in more detail in the preceding chapter, and sparked responses that included provision of food and recipes, and home visits. Material support for learning was also needed by many students, with an eye to all elements of the school curriculum.

> [We] included things like for PE, we put a skipping rope and a ball and a list of activities. We did visual art, we sent paint home. So we put in a little container like a soy sauce container, some paint, paint brushes. We actually had coloured pencils, textas, crayons, scissors, rulers, pens, pencils, everything we bought. (School-based leader as cited in Brown et al., 2020, p. 43)

2. *Any use of technology mediated learning must take into account digital inclusion*

   The term "digital inclusion" has arisen in response to a growing digital economy, and the imperative to support all members of the community to have access to online participation (Thomas et al., 2020). In Australia, a Digital Inclusion Index report for the nation is produced annually. Data are collected against demographic characteristics in terms of three elements: Access (to internet, devices, and data); Affordability; and Digital Ability. From this Index, there has been a clearly identified gap in digital inclusion on economic and social grounds. Gaps in inclusion in the first two components of the Digital Inclusion Index, access (to devices, internet, and data) and affordability, largely reflect and are impacted by disparities in economic capital (Bourdieu, 1984). The third component, digital ability, can be considered in relation to cultural capital (Bourdieu, 1984), encompassing attitudes, skills, and capabilities. Digital ability is relevant to both students and parents, particularly for younger children learning at home:

> Parents are very grateful for the support whether it is the loan of devices and/or data cards. However, they are finding this a very difficult time because they feel obliged to help the students and they are unable to do so. (Secondary school-based staff as cited in Brown et al., 2020, p. 49)

There are key groups who are challenged in terms of digital capability; those with low literacy skills, or who have had little or no opportunity to develop technology skills (Thomas et al., 2020). Geography also plays a part (and is also linked with a family's economic capital), with rural and remote families particularly challenged (Thomas et al., 2020). This type of gap is also found in the international context. In some developing countries, the issue goes beyond the digital divide, but also encompasses reliable electricity and adequate housing and the compounding effects of living in poverty (Okebukola et al., 2020). Indeed, even in the Australian context, access to adequate housing (Robinson, 2020) and electricity was an issue for some students, depriving them of the option of online learning as an education system officer in our research explained: "a lot of the students don't actually have any form of power in the house, so they're living in squalid conditions so that makes it incredibly difficult." When deficiency needs such as these go unmet for vulnerable children and young people (Noltemeyer et al., 2021), clearly, digital inclusion is also compromised, though in such instances, other material needs may be of greater urgency.

3. *Traditional technologies can be used to deliver content but have limitations*

   The use of television or radio to deliver educational content to students has several advantages beyond addressing the digital divide. As familiar technologies, their usage was inclusive of families without access to a computer or mobile device, or those who lacked confidence, experience, or digital literacy skills to productively engage with the online content. However, there are limitations with the use of broadcasting technologies. They are difficult to tailor to the diversity of students, and the delivery is one-way, with limitations on interactivity. Beyond an emergency response, they would require supplementation to ensure key aspects of the learning environment are made available to students. Although the OECD (2021) identified that countries using television and radio also provided professional learning and support to teachers who were employing these media, the efficacy of the approach in the longer term is difficult to evaluate in the absence of contextualised data (Dabrowski et al., 2020).

4. *Online learning provides opportunities that can engage, support, and extend learners*

   Asynchronous online learning, accessible anytime—together with the use of a diversity of resources—was seen by many as empowering students to take more control of their learning, as a secondary school teacher in our research revealed.

   > …A lot of kids decided they'd like to log in at 8pm at night, that was when they could learn really well. They could pace themselves slowly if they really wanted to go deeper with the content. (School-based staff)

   There were also specific groups of students who responded well to the online environment, including those who had previously disengaged from school:

   > I think online learning is something that all schools should embrace as part of their Tier 3 programs[4] to re-engage children that may have disengaged from education for other reasons. (School-based staff as cited in Brown et al., 2020, p. 53)

   Online learning does not need to be abandoned with a return to school. The inroads made through a period of rapid adaptation and understanding what elements of online practice are effective can be used to supplement in class programs.

5. *Adopting technology-enhanced learning, or blended learning, requires significant attention to purposeful design and student readiness*

   Employing online learning during the pandemic provided evidence that simply providing content through technology, no matter how engaging, or of what quality, is not sufficient to promote learning. This understanding is well progressed in higher education, and by other quality e-learning providers, underpinned by theories of learning and learning design. Social constructivism and connectivism ideas provide the foundation for learning design that centres on the

[4] In the state of Tasmania in Australia, Tier 3 programs are for learners "identified as being at significant risk of disengaging", usually offered as a separate program on the school site (Tasmanian Department of Education, 2021).

student, and necessarily incorporates interactivity, feedback, and positive and purposeful engagement with peers (e.g., Parker et al., 2013; Siemens, 2005). However, the road towards online and blended learning in the higher education sector has been long and contested and continues to be the subject of action research projects for improvement. Applying this learning environment to younger students brings with it added complexity.

It may be argued that the disruption and associated rapid adoption through necessity has driven uptake of technology enhanced implementation in ways that were not possible without such an imperative. Instructor readiness has been identified as a key challenge in the adoption of online teaching (e.g., Okebukola et al., 2020). The speed of the adaptation to learning at home mitigated against having the necessary time available to fully, and purposefully, design an e-learning program. Nevertheless, the accelerated attention to aspects of design, systemic attention to just-in-time professional and peer to peer learning and system-wide initiatives made significant progress on effective online learning design for school students.

Another critical enabler is ensuring student readiness. In the case of very young people, this also extends to parents and caregivers. The assumption that today's young people are "digital natives" and confident and capable in the online environment does not necessarily hold. A study of Finnish secondary students found attitudes and beliefs can affect readiness for, and preparedness to engage in, online learning (Valtonen et al., 2009). In addition, the need for skills in independent learning to cope in the online environment was challenging for some students and was identified amongst school-based staff in our research: "Not many of these young people are able to work independently" (School-based staff as cited in Brown et al., 2020 p. 50).

6. *Parents and caregivers can be key partners but may require support, upskilling, and time*

   Time that parents and caregivers spend with their children on productive or educational activities is important to a child's development (Attanasio et al., 2020; Del Bono et al., 2016; Fiorini & Keane, 2014). However, with a shift to learning at home, the role expanded to include additional responsibilities. Whilst some may be balanced with other tasks, parents and caregivers cannot be expected to replicate the roles and functions of teachers or provide the infrastructure and resources available in schools within the home.

Being mindful of the range of responsibilities in families provides a foundation for understanding the limitations of what is asked, and the support that might be given. The inability to assist their children with technology has also been found to be a source of anxiety (Goedhart et al., 2019). Having open discussions with families and including families in planning for learning can also serve to address any concerns or attitudes that may affect families' adoption of online, or blended, learning. This may include debunking myths that delivering content in an online format necessarily requires sitting in front of a device. Goedhart et al. (2019) uncovered a commonly held belief of mothers that "children below the age of five should not have access to ICT, explaining that it is 'bad for their eyes' and 'bad for their brains'" (p. 2355). As noted in Chaps. 4 and 5, in addition to having skills and knowledge, parents and carers also need time. When juggling employment and household duties, it can be difficult to find time to engage productively with children's formal (school) and nonformal learning. In this respect, families that can afford to have one parent/carer full-time at home while the other is on paid employment have an advantage.

Increasing confidence with technology as a mode for both learning and communication has opened up new ways for teachers and schools to communicate with parents and caregivers. Used effectively as one secondary school teacher explained, it can demystify curriculum and planning for learning, involving students and families more closely in the learning process:

> If you're going to do the work and put time into planning, you might as well have it in a space where kids can access it 24/7, and where kids can interact with it and revisit the content. (Secondary school-based staff)

How to Move Forward?

The key focus of this chapter is to highlight how we can use the experiences of the pandemic to develop new ways of learning for young people. In doing so, we will highlight two key emergent themes. The first, through the affordances of technology to enhance learning; the second, to strengthen new ways of building home-school partnerships to promote learning.

Technology Enhanced Learning

Even prior to the events of 2020, educational researchers were identifying the potential of online learning for extension of learning for high achievers; supplementation and individualisation for students experiencing challenges or who had missed school; or for providing increased options for students in regional or remote communities (e.g., Saultz & Fusarelli, 2017). Schools and systems were increasing their adoption of these methods, and there has been a growth in open, and for-profit educational websites (e.g., Kahn Academy, WooTube). However, fully online, or virtual schools have had mixed success in terms of results for students (Saultz & Fusarelli, 2017).

Employed wisely, technology has the potential to enrich learning experiences through high quality and engaging resources and the facilitation of interaction between students and teachers and peer to peer. This is combined with a flexibility for students with respect to time and place of study (Brown et al., 2013).

The open educational resource (OER) movement has contributed to the potential benefits of online learning through opening access to quality resources free of charge, but also by catalysing a community that share, review, remix, enhance and reuse online content in a collaborative way (McGreal, 2017). Modern learning analytics are also increasingly providing a personalised learning experience for students (see Tan et al., 2016). However, creation of engaging, high quality online learning requires time, and digital skills on behalf of both the teachers and the students.

Purposeful design underpins effective learning experience, regardless of delivery mode. There are a number of key heuristics and taxonomies that are used in designing learning experiences, that can be applied across contexts and media (see Anderson & Krathwohl, 2001; Biggs, 1996; Bloom et al., 1956; Shulman, 2002). However inherent in the use of these frameworks is a consideration of the learner and the context. It stands to reason, therefore, that a change in context—even if the intended learning is the same—requires a redesign of the learning experience.

Drawing from what we have learnt from the use of an online environment during the COVID-19 pandemic, Fig. 6.2 represents key elements that should be considered in any such redesign.

Online learning in the compulsory school years should adopt a nuanced model that reflects the stage of development of the students. Guided by the curriculum goals, curated, and scaffolded content can be sequenced

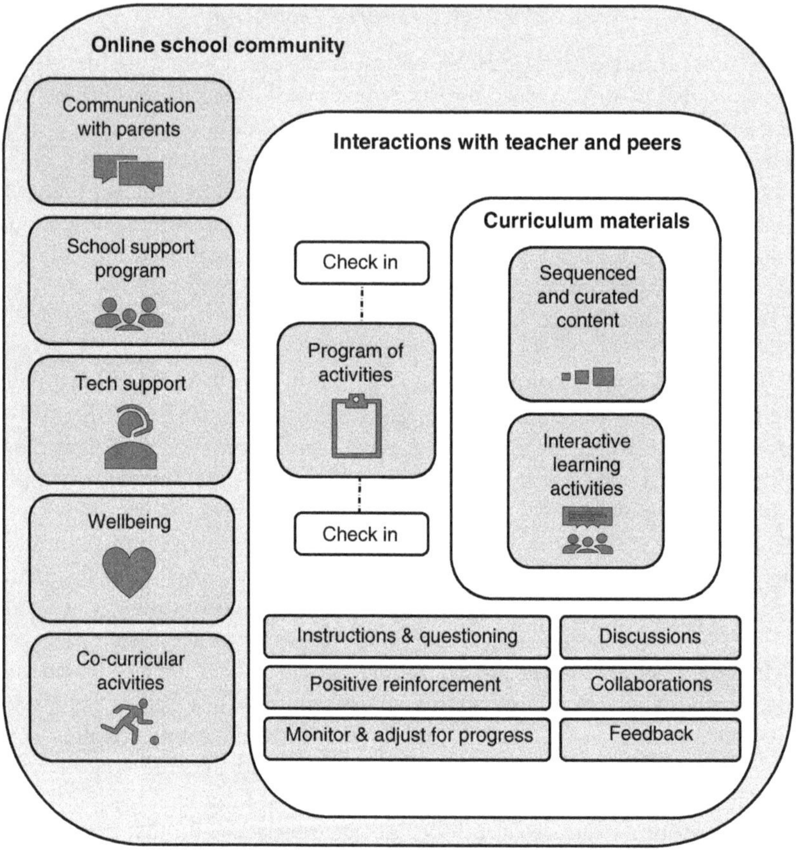

Fig. 6.2 Model demonstrating an example of an online school community. Figure adapted from Brown et al. (2020, p. 51)

supported by activities that encompass the elements essential to good pedagogy (instruction and questioning, positive reinforcement and feedback, an ability to monitor and adjust appropriately for progress, and the opportunity to explore areas of interest; engage in practice, and discuss or collaborate with teachers and peers).

Building a sense of community should be an explicit component of any online environment, with better performance at school for school-aged young people linked to socio-emotional connections with teachers and peers (Darling-Hammond et al., 2020). This interaction can, and should, occur in a combination of real time and asynchronous modes. A learning management system or an enterprise communication system can provide a secure environment for this to take place.

To create a holistic learning environment, incorporation of wrap-around supports into the online learning environment that connect students (and families) with support services and technical support should be seamlessly integrated. Whilst the representation in Fig. 6.2 outlines the components of an online learning environment, it does not tell the whole story. Issues such as digital inclusion become increasingly critical to dialogue and policy surrounding learning, particularly online. Alongside a purposeful design for learning, the importance of structure in a learning program has also been identified. In their recent study in the Netherlands, Lauret and Bayram-Jacobs (2021) investigated the provision of online learning with respect to eight key foundations of instructional practice. They note student feedback on online learning has clearly identified the importance of a clear and explicit structure. Instructional design, and specialist instructional designers as change agents, well-established in Higher Education (Campbell et al., 2009) may, therefore, have a more prominent place in planning for learning in the school sector.

This disruption to status quo has provided an impetus for change that has implications into the future. Expanding the pedagogical repertoire of teachers to include online delivery has increased possibilities of individualised programs, flexibility, collaborations beyond the classroom and access to a broader range of resources to engage students in learning. The rapidity of the move resulted in a significant degree of collaboration between teachers, and within school systems to facilitate development and delivery of curriculum online. This also had positive impacts on the way planning for learning happened in schools. The benefits for both teachers and students suggest changes in practice that could be constructive going forward, as a secondary school teacher in our study explained:

> … we just made a shift to all of our planning is going to happen in Canvas [learning management system]. That way, we've all got access to it, the kids have got access to it, the parents can see it. It's de-privatising practice. (Secondary school-based staff)

The affordances of online learning, post-crisis, do need to be considered with a return to classrooms. How can this learning be translated into a blended space? Ideally, the elements of any learning design model could be positioned across the physical and virtual spaces, with neither one privileged over the other in the broadest sense, but very much tailored to the student and the family at the individual level.

Specifically in relation to disadvantaged students in the context of the pandemic, online technology has provided opportunities for additional tutoring to help these students catch up following the loss of instructional time due to school closures. In Australia, recommendations to invest in a "catch-up strategy …for disadvantaged students hardest hit by the shift to remote schooling during the COVID-19 crisis" (Sonnemann & Goss, 2020, p. 4) were taken up by some state governments and NGOs. Importantly, these programs involved tutors working with students online, making tutoring more widely available (New South Wales Auditor General, 2021; The Smith Family, 2021). The evaluation of one program noted online delivery as a key enabler:

> The online delivery of Catch-Up Learning enhanced the reach of the program, giving access to students who may otherwise have been excluded because of where they lived. It also enabled tutors to be best matched to students and made it easier to re-schedule lessons when required because of the complexity of families' lives and tutor availability. (The Smith Family, 2021, p. 11)

Importantly, this program also provided technology support, such as hardware and internet connection for families, and online education programs and assessment tools for tutors.

A New School: Home Dynamic

A key aspect of the shift to a shared responsibility between school and home, became a focus on school-home partnerships. Parents/caregivers have long been recognised as children's first teachers, and as essential partners for achievement in school learning regardless of how or where learning is delivered (Epstein, 1987). Indeed, the positive impact of parental involvement has been identified as a way to "mediate the effects of poverty, parents' educational attainment, and race/ethnicity on achievement" (Lee & Bowen, 2006, p. 194).

Despite the potential, there are recognised challenges to fostering parental involvement with school, particularly in the latter years of schooling. The reasons are complex and encompass time constraints on working parents; location issues for regional or remote parents; as well as the effects of differing levels of social capital amongst parents that influence how the school relationship is navigated (Antony-Newman, 2018; Bourdieu, 1984). The differential engagement of parents with their child's schooling contribute to the persistence of educational inequalities.

During the COVID-19 pandemic, the increase in parent/caregiver awareness of, and involvement in, school curriculum through learning at home introduced new ways of creating school-home partnerships for learning. Insights into the complexity of the elements of the teaching program beyond "content delivery" have provided a platform for deeper understanding (and appreciation) of the education process. The use of mobile technology, telephone calls, learning management systems, online conferencing to support learning at home, also increased the contact with parents and caregivers.

Some parents have gained a greater insight into the school curriculum, the importance of aspects of classroom teaching (such as feedback, engagement with peers and creating a sense of belonging), and ways in which they can feed into the learning process to benefit their students (Australian Curriculum, Assessment and Reporting Authority, 2021). For some parents the opportunity to engage in parent-teacher interviews remotely has meant the first formal contact with the school.

The establishment of new ways of supporting parents/caregivers to facilitate and encourage learning should not be lost as students return to the schooling environment. Positive school-home dynamics and parental involvement are both areas identified as successful interventions in education by Hattie (2015). Building or maintaining strong school-home partnerships, especially for vulnerable or disadvantaged students have clear potential benefits. However, to capitalise on this, issues of access, affordability and confidence with any technologies used need to be addressed. The issue of digital inclusion is therefore critical (Thomas et al., 2020) and was reinforced through the comments of a primary school teacher in our research:

> Seesaw and Canvas [online learning tools] allowed teaching staff to share information quickly and easily with students and families; many families of younger students were already working with Seesaw before the learning at home period, ensuring families were comfortable and familiar with the tool. (Primary school-based staff)

For many parents, internet provision for educational purposes has gained greater relevance and importance and prompted activism such as that of the Isolated Children's Parent's Association (ICPA) in Australia. This group is calling on governments to improve internet provision, and for subsidisation of data for educational purposes (ICPA, 2021).

Although clearly providing opportunity, there does need to be a word of caution in the adoption of technology concerning issues broadly related to cyber security. A recent issue arose in the largest Australian jurisdiction where deactivation of communications and learning management system was required (Chrysanthos, 2021). The adoption of learning management systems, especially for senior secondary students, increases vulnerability of students to such attacks, and requires resourcing of ICT professionals to ensure security of data and service.

Increased attention to digital literacy and awareness of surveillance and privacy concerns (Bailey et al., 2020) also need to be considered when choosing, implementing, and supporting digital platforms for educational purposes. These issues are best dealt with at school, or even system, level to provide systems and guidelines for teaching practice. Attention to this important aspect requires resourcing to introduce and implement and necessarily involves school, teachers, students, and parents/caregivers (Laskowski, 2020). However, this does represent an opportunity for schools to increase digital awareness and digital literacy of families and broader school communities.

Conclusion

There are compelling reasons to continue to develop a blended learning approach that draws on the experiences of the rapid shift to learning at home during the pandemic. Incorporating essential elements that have been identified through learning at home is already well progressed in many schools and systems. Continuing to develop this, together with focussed professional development, and time for teachers to design complementary online and blended experiences has the potential to enhance and personalise curriculum delivery for a broad range of students, whether learning at school, or at home.

The expanded use of technology has a range of potential benefits to increase the connection between school and home through greater transparency and accessibility to education programs for parents and caregivers. This represents an unparalleled opportunity to build capability in schools and families to build positive collaborations to promote educational opportunity.

To fully capitalise on this potential, issues of digital inclusion need to be addressed through systematic attention by policy makers at all levels of government, and in education systems. This must go beyond provision of infrastructure and tackle issues of affordability and capability. Although the pandemic has provided an impetus for change, continued attention and resources must be directed towards change management, and purposeful design of the renewed educational ecosystem.

References

Adnan, M., & Anwar, K. (2020). Online learning amid the COVID-19 pandemic: Students' perspectives. *Journal of Pedagogical Sociology and Psychology, 2*(1), 45–51. https://doi.org/10.33902/JPSP.2020261309

Altmann, E., Kasthuri, S., & Jose, H. (2014). *Perspective: Any content, anytime, anywhere, any device.* Infosys. Retrieved July 8, 2021, from https://www.infosysbpm.com/offerings/industries/media-entertainment/white-papers/documents/any-content-anytime-anywhere-device.pdf

Anderson, L. W., & Krathwohl, D. R. (Eds.). (2001). *A taxonomy for learning, teaching and assessment: A revision of Bloom's taxonomy of educational objectives.* Pearson Education.

Antony-Newman, M. (2018). Whose involvement has capital? The application of Bourdieu's thinking tools for the study of parental involvement. *Social Theory Applied.* Retrieved August 2, 2021, from https://socialtheoryapplied.com/2018/12/17/8842/

Attanasio, O., Blundell, R., Conti, G., & Mason, G. (2020). Inequality in socio-emotional skills: A cross-cohort comparison. *Journal of Public Economics, 191*, 104171. https://doi.org/10.1016/j.jpubeco.2020.104171

Australian Broadband Advisory Council. (2020). *Riding the digital wave report on COVID-19.* Retrieved May 6, 2021, from https://apo.org.au/sites/default/files/resource-files/2020-11/apo-nid309733.pdf

Australian Curriculum, Assessment and Reporting Authority. (2021). *The COVID-19 education impact—Are parents more engaged?* Retrieved March 25, 2021, from https://www.acara.edu.au/docs/default-source/media-releases/20210322-covid-research-media-release.pdf

Australian Institute of Health and Welfare. (2021). *Australia's youth—In brief.* (Cat. No. CWS 80). Retrieved July 1, 2021, from https://www.aihw.gov.au/reports/children-youth/australias-youth-in-brief/contents/summary

Bailey, J., Burkell, J., Regan, P., & Steeves, V. (2020, April 22). Children's privacy is at risk with rapid shifts to online schooling under coronavirus. *The*

Conversation. Retrieved February 15, 2021, from https://theconversation.com/childrens-privacy-is-at-risk-with-rapid-shifts-to-online-schooling-under-coronavirus-135787

Balanzategui, J., Burke, L., & McIntyre, J. (2021). 'What would Bandit do?': Reaffirming the educational role of Australian children's television during the COVID-19 pandemic and beyond. *Media International Australia, 178*(1), 54–62. https://doi.org/10.1177/1329878X20948272

Basilaia, G., & Kvavadze, D. (2020). Transition to online education in schools during SARS-CoV-2 coronavirus (Covid-19) pandemic in Georgia. *Pedagogical Research, 5*(4), em0060. https://doi.org/10.29333/pr/7937

Biggs, J. (1996). Enhancing teaching through constructive alignment. *Higher Education, 32*, 347–364. https://doi.org/10.1007/BF00138871

Bloom, B. S., Englehart, M. D., Furst, E. J., Hill, W. H., & Krathwohl, D. R. (1956). *The Taxonomy of educational objectives, handbook I: The cognitive domain*. David McKay Co., Inc..

Bourdieu, P. (1984). *Distinction: A social critique of the judgement of taste* (R. Nice, Trans.). Harvard University Press.

Brown, N., Kregor, G., Williams, G., Padgett, L., Bossu, C., Warren, V., & Osborne, J. (2013). *Technology enhanced learning and teaching white paper 2014–2018*. Tasmanian Institute of Learning and Teaching, University of Tasmania. Retrieved November 14, 2020, from https://www.teaching-learning.utas.edu.au/__data/assets/pdf_file/0003/439014/Technology-Enhanced-Learning-and-Teaching-White-Paper_Background-papers-Academic-Senate-15-November-2013.pdf

Brown, N., Riele, K. T., Shelley, B., & Woodroffe, J. (2020). *Learning at home during COVID-19: Effects on vulnerable young Australians* (Independent Rapid Response Report). Peter Underwood Centre for Educational Attainment, University of Tasmania. Retrieved May 26, 2020, from https://www.utas.edu.au/__data/assets/pdf_file/0008/1324268/Learning-at-home-during-COVID-19-updated.pdf

Bubb, S., & Jones, M.-A. (2020). Learning from the COVID-19 home-schooling experience: Listening to pupils, parents/carers and teachers. *Improving Schools, 23*(3), 209–222.

Campbell, K., Schwier, R. A., & Kenny, R. F. (2009). The critical, relational practice of instructional design in higher education: An emerging model of change agency. *Educational Technology Research and Development, 57*(5), 645–663.

Chrysanthos, N. (2021, July 8). NSW teachers in 'state of paralysis' after cyber attack. *The Sydney Morning Herald*. Retrieved July 8, 2021, from https://www.smh.com.au/national/nsw/nsw-teachers-in-state-of-paralysis-after-cyber-attack-20210708-p5883b.html

Cousins, S. (2020, July 8). Children in this Australian town don't have the internet, so their school has turned to radio. *SBS News*. Retrieved July 8, 2021, from https://www.sbs.com.au/news/children-in-this-australian-town-don-t-have-the-internet-so-their-school-has-turned-to-radio

Coyne, L. W., Gould, E. R., Grimaldi, M., Wilson, K. G., Baffuto, G., & Biglan, A. (2020, April 12). First things first: Parent psychological flexibility and self-compassion during COVID-19. *Behavior Analysis in Practice*. https://doi.org/10.1007/s40617-020-00435-w

Cullinane, C., & Montacute, R. (2020). *COVID-19 and social mobility impact brief #1: School shutdown*. The Sutton Trust. Retrieved August 15, 2020, from https://www.suttontrust.com/wp-content/uploads/2021/01/School-Shutdown-Covid-19.pdf

Dabrowski, A., Nietschke, Y., Taylor-Guy, P., & Chase, A. (2020). *Mitigating the impacts of COVID-19: Lessons from Australia in remote education*. Australian Council for Educational Research. https://research.acer.edu.au/learning_processes/32/

Darling-Hammond, L., Schachner, A., & Edgerton, A. K. (2020). *Restarting and reinventing school: Learning in the time of COVID and beyond*. Learning Policy Institute. http://learningpolicyinstitute.org/product/restarting-reinventing-school-covid

Del Bono, E., Francesconi, M., Kelly, Y., & Sacker, A. (2016). Early maternal time investment and early child outcomes. *Economic Journal, 126*(596), F96–F135. https://doi.org/10.1111/ecoj.12342

Dickinson, H., & Yates, S. (2020). *More than isolated: The experience of children and young people with disability and their families during the COVID-19 pandemic*. Children and Young People with Disability, Australia. Retrieved January 19, 2021, from https://www.cyda.org.au/images/pdf/covid_report_compressed_1.pdf

Epstein, J. L. (1987). Parent involvement: What research says to administrators. *Education and Urban Society, 19*(2), 119–136. https://doi.org/10.1177/0013124587019002002

Eurochild. (2020). *Growing up in lockdown: Europe's children in the age of COVID-19* (2020 Eurochild Report). Retrieved July 15, 2021, from https://www.eurochild.org/resource/growing-up-in-lockdown-europes-children-in-the-age-of-covid-19/

Fay, J., Levinson, M., Stevens, A., Brighouse, H., Geron, T., & Fellow, P. (2020). *Schools during the COVID-19 pandemic: Sites and sources of community resilience* (White Paper 20). Edmond J. Safra Center for Ethics, Harvard University. Retrieved November 20, 2020, from https://ethics.harvard.edu/files/center-for-ethics/files/20schoolsduringpandemic2.pdf?m=1592232939

Fiorini, M., & Keane, M. P. (2014). How the allocation of children's time affects cognitive and noncognitive development. *Journal of Labor Economics, 32*(4), 787–836. https://doi.org/10.1086/677232

Giuffre, L. (2021). Bluey, Requestival, Play School and ME@Home: the ABC (Kids) of communication cultures during lockdown. *Media International Australia, 178*(1), 63–76. https://doi.org/10.1177/1329878X20952520

Goedhart, N., Broerse, J., Kattouw, R., & Dedding, C. (2019). 'Just having a computer doesn't make sense': The digital divide from the perspective of mothers with a low socioeconomic position. *New Media and Society, 21*(11–12), 2347–2365. https://doi.org/10.1177/1461444819846059

Graham, A., & Sahlberg, P. (2020, March 26). Schools are moving online but not all children start out digitally equal. *The Conversation*. Retrieved March 27, 2020, from https://theconversation.com/schools-are-moving-online-but-not-all-children-start-out-digitally-equal-134650

Harrington, M. (2011). *Australian Government funding for schools explained.* Parliament of Australia. Retrieved July 8, 2021, from https://www.aph.gov.au/About_Parliament/Parliamentary_Departments/Parliamentary_Library/pubs/BN/1011/SchoolsFunding#_Toc284226207

Hattie, J. (2015). The applicability of visible learning to higher education. *Scholarship of Teaching and Learning in Psychology, 1*(1), 79–91. https://doi.org/10.1037/stl0000021

Isolated Children's Parents' Association. (2021). *ICPA Briefing Papers March 2021*. Retrieved July 8, 2021, from https://icpa.com.au/sites/default/files/inline-files/ICPA%20Aust%20Full%20Briefing%20Papers%20-%20March%202021.pdf

Laskowski, T. (2020). *Cybersecurity guidelines for remote learning. ASCD, A new reality: Getting it right. Educational Leadership Special Report, 77*(10). Retrieved February 16, 2021, from https://www.ascd.org/el/articles/cybersecurity-guidelines-for-remote-learning

Lauret, D., & Bayram-Jacobs, D. (2021). COVID-19 lockdown education: The importance of structure in a suddenly changed learning environment. *Education Sciences, 11*, 221. https://doi.org/10.3390/educsci11050221

Lee, J.-S., & Bowen, N. K. (2006). Parent involvement, cultural capital, and the achievement gap among elementary school children. *American Educational Research Journal, 43*(2), 193–218. https://doi.org/10.3102/00028312043002193

Lee, V., & Yeo, S. (2020, April 18). How home-based learning shows up inequality in Singapore—A look at three homes. *The Straits Times*. Retrieved July 5, 2021, from https://www.straitstimes.com/lifestyle/how-home-based-learning-hbl-shows-up-inequality-in-singapore-a-look-at-three-homes

McDaid, L., Tomaszewski, W., Staton, S., Povey, J., Thorpe, K., Vaithianathan, R., Ridgway, T., Diffey, B., & Western, M. (2020). *Who are the students most at risk of falling being in their learning?* (Pillar Report No. 1). Institute for Social Research, University of Queensland. Retrieved August 19, 2021, from https://issr.uq.edu.au/files/17185/Ramsay%20Pillar%201_Full%20Report.pdf

McGreal, R. (2017). Special report on the role of open educational resources in supporting the sustainable development goal 4: Quality education challenges and opportunities. *International Review of Research in Open and Distributed Learning, 18*(7), 292–305. https://doi.org/10.19173/irrodl.v18i7.3541

Mhlanga, D., & Moloi, T. (2020). COVID-19 and the digital transformation of education: What are we learning on 4IR in South Africa? *Education Sciences, 10*(7), 180. https://doi.org/10.3390/educsci10070180

New South Wales Auditor-General. (2021). *COVID intensive learning support program. Performance audit.* https://www.audit.nsw.gov.au/our-work/reports/covid-intensive-learning-support-program

Noltemeyer, A., James, A. G., Bush, K., Bergen, D., Barrios, V., & Patton, J. (2021). The relationship between deficiency needs and growth needs: The continuing investigation of Maslow's theory. *Child & Youth Services, 42*(1), 24–42.

Öçal, T., Halmatov, M., & Alta, S. (2021). Distance education in COVID-19 pandemic: An evaluation of parent's, child's and teacher's competencies. *Education and Information Technologies, 26*, 6901–6921. https://doi.org/10.1007/s10639-021-10551-x

Okebukola, P. A., Suwadu, B., Oladejo, A., Nyandwi, R., Ademola, I., Okorie, H., & Awaah, F. (2020). Delivering high school chemistry during COVID-19 lockdown: Voices from Africa. *Journal of Chemical Education, 97*, 3285–3289. https://doi.org/10.1021/acs.jchemed.0c00725

Organisation for Economic Co-operation and Development. (2021). *The State of education one year into the COVID pandemic.* https://doi.org/10.1787/201dde84-en

Parker, J., Major, D., & Herrington, J. (2013). Authentic online learning: Aligning learner needs, pedagogy and technology. *Issues in Educational Research, 23*(2), 227–241.

Reimers, F., & Schleicher, A. (2020). *A framework to guide an education response to the COVID-19 pandemic of 2020.* OECD Publishing. https://doi.org/10.1787/6ae21003-en

ReliefWeb. (2020). H*ow countries are using edtech (including online learning, radio, television, texting) to support access to remote learning during the COVID-19 pandemic.* Retrieved May 5, 2021, from https://reliefweb.int/report/austria/how-countries-are-using-edtech-including-online-learning-radio-television-texting

Robinson, C. (2020). *#StayHome? The impact of COVID-19 on unaccompanied homeless children in Tasmania.* Anglicare. Tasmania. Retrieved October 20, 2021, from https://www.anglicare-tas.org.au/research/stayhome-a-full-report-of-the-impact-of-covid-19-on-unaccompanied-homeless-children-in-tasmania/

Sahlberg, P. (2021). Does the pandemic help us make education more equitable? *Education Research Policy and Practice*, *20*, 11–18. https://doi.org/10.1007/s10671-020-09284-4

Saultz, A., & Fusarelli, L. D. (2017). Online schooling: A cautionary tale. *Journal of School Choice*, *11*(1), 29–41. https://doi.org/10.1080/15582159.2016.1272928

Shulman, L. S. (2002). Making differences: A table of learning. *Change: The Magazine of Higher Learning*, *34*(6), 36–44. https://doi.org/10.1080/00091380209605567

Siemens, G. (2005). Connectivism: A learning theory for the digital age. *International Journal of Instructional Technology and Distance Learning*, *2*(1). http://www.itdl.org/

Sonnemann, J., & Goss, P. (2020). *COVID catch-up: Helping disadvantaged students close the equity gap*. https://grattan.edu.au/report/covid-catch-up/

Tan, Q., Zhang, X., Pivot, F., Evans, J., Kinshuk, D., & McGreal, R. (2016). The 5R adaptation framework: Concepts, systems, and learning scenarios. *Journal of Internet Technology*, *17*, 971–980. https://doi.org/10.6138/JIT.2016.17.5.20141119a

The Smith Family. (2021). *Improving young Australians' literacy and numeracy. The catch-up learning program*. https://www.thesmithfamily.com.au/research/reports/catch-up-learning-pilot

Thomas, J., Barraket, J., Wilson, C. K., Holcombe-James, I., Kennedy, J., Rennie, E., Ewing, S., & MacDonald, T. (2020). *Measuring Australia's digital divide: The Australian digital inclusion index 2020*. Centre for Social Impact, Royal Melbourne Institute of Technology and Telstra. https://doi.org/10.25916/5f6eb9949c832

United Nations. (2020). *Policy Brief: Education during COVID-19 and beyond*. Retrieved April 4, 2021, from https://www.un.org/development/desa/dspd/wp-content/uploads/sites/22/2020/08/sg_policy_brief_covid-19_and_education_august_2020.pdf

United Nations Educational, Scientific and Cultural Organization. (2020). *COVID-19 educational disruption and response*. Retrieved April 4, 2021, from https://en.unesco.org/covid19/educationresponse

Valtonen, T., Kukkonen, J., Dillon, P., & Väisänen, P. (2009). Finnish high school students' readiness to adopt online learning: Questioning the assumptions. *Computers & Education*, *53*, 742–748. https://doi.org/10.1016/j.compedu.2009.04.014

Zhao, Y., & Watterson, J. (2021). The changes we need: Education post COVID-19. *Journal of Educational Change*, *22*, 3–12. https://doi.org/10.1007/s10833-021-09417-3

CHAPTER 7

Educational Equity in Times of Crisis

Abstract This chapter gathers the findings from our research into education during the COVID-19 pandemic with lessons from our examination of how prior disasters and crises have affected educational equity for children and young people. This chapter summarises our three key concerns for children and young people pursuing education in times of crisis: material needs, emotional wellbeing, and access to learning, and introduces our model of educational equity. We draw attention to the impacts of structural disadvantage and underscore the importance of learning, engagement, and support in navigating in uncertain times. A key takeaway is that disasters and crises exacerbate inequities, but also spur opportunity for innovation in education.

Keywords COVID-19 pandemic • Material needs • Emotional wellbeing • Access to learning • Model of educational equity

Introduction

The COVID-19 pandemic has presented both a global shock to education systems as well as an opportunity to radically reconsider what educational equity means in times of crisis and disaster. In such times, attending to educational equity is essential to support all children and young people to have opportunities to live meaningful and fulfilling lives.

E. S. Rudling et al., *Education and Equity in Times of Crisis*, https://doi.org/10.1007/978-3-031-18671-4_7

Crises compound the pressure on schools to care for all learners. However, historical crises teach us that emergencies tend to exacerbate inequities, as well as deepen the reliance of marginalised students on the emotional, material, and learning supports provided by school. The COVID-19 pandemic has been the focal point in this book, through which we explored what can be learned about the role of schools during and beyond crisis—and about implications for educational equity more broadly.

The concept of vulnerability is a key theme in this book (see especially Chap. 2). We view vulnerability as a product of broader socio-economic structures and systems. In this respect, vulnerability predates the COVID-19 pandemic, and children's and young people's experiences of such events are framed by pre-existing vulnerabilities. As in historical disasters and crises, the effects of the pandemic have reiterated how such events can also create cohorts of newly vulnerable students.

Disaster and crisis affect many people in different ways. People in positions of privilege may also find their home damaged by disaster (flood, earthquake, bushfire) and struggle with school closures, but they have access or resources (economic, cultural, and social capital) that assist in mitigating the impact of such challenges. The effects of disasters and pandemics are unevenly distributed—the disproportionate impacts are reflected across socio-economic divides. Those already experiencing disadvantage, including intergenerational and entrenched disadvantage, are likely to be hardest hit during a crisis.

This is, in part, because crises tend to happen suddenly, with little time to prepare. As examined in Chap. 3, previous disasters show how schools can be destroyed by tsunamis and hurricanes or how viral pandemics lead authorities to quickly close schools to reduce virus transmission. However, despite there being little time to prepare for crises, analysis of historical events presents opportunities to establish preventative measures that limit the impact of disasters and pandemics on educational equity.

Across the book, to capture the complexity of vulnerability and educational equity, we have drawn on a suite of conceptual tools (see Chap. 1). Bronfenbrenner's ecological systems theory (1995) helps to identify the impacts of crisis on educational outcomes, as influenced by different layers of the education ecosystem. Theories of capital (Bourdieu, 1984; Putnam, 2000) make visible the influence of family, social and cultural capital on student learning and wellbeing. Contributions by Maslow (1954) and Noltemeyer et al. (2021) afford insight into the nature of students' deficiency needs and growth needs.

Applying this conceptual framework to our research, we conclude there are three key concerns for educational equity in times of crisis:

- material needs,
- emotional wellbeing, and
- access to learning.

Figure 7.1 represents the model we have developed through our analysis and which we discuss in this final chapter.

Material needs, emotional wellbeing, and access to learning are linked with, and mediated by, family, learning contexts, and education systems, as

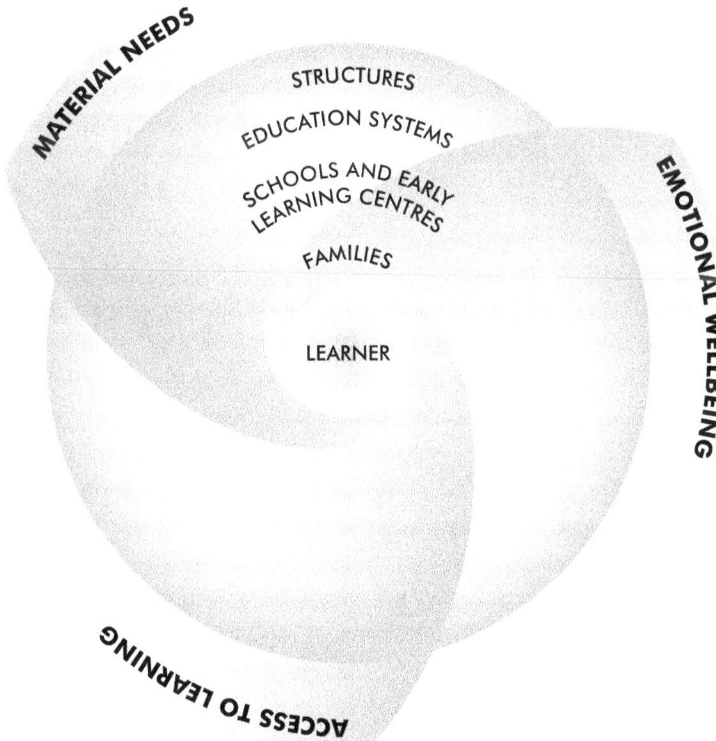

Fig. 7.1 Model of educational equity for learners. Figure adapted from Emery (2019, p. 139)

well the structural dimensions that shape everyday life. Following Bronfenbrenner (1995), we put the learner in the middle of this model since they are the primary beneficiary of education. In order to enhance educational equity during times of crisis, attention needs to be paid to each layer and consideration given to their interdependencies. This model may also prove fruitful for working towards educational equity more widely—outside of responses to crises.

Our model captures the complexity of vulnerability, and the susceptibility of already disadvantaged children and young people to educational inequities, when crisis strikes. While the specific pressures that crises create will vary, across society all dimensions and layers form a web of interconnected factors that affect educational equity. For convenience, we address each dimension in turn in the sections below; however, in actual responses, they cannot be treated in isolation.

Our conclusions in this chapter are driven by findings from our own research with a wide range of school staff and other experts, as well as our extensive examination of literature from prior crises and—especially—in relation to education during the COVID-19 pandemic. Our key takeaway is that disaster and crisis exacerbate inequities, but also spur opportunity for innovation. Chapters 2–6 contain the detailed evidence that sits behind the conclusions we present here.

We acknowledge that educational inequity is patterned similarly to inequities in society more broadly. This book has identified constraints and enabling initiatives at different layers of the education ecosystem, but ultimately educational equity will be served best by a more equitable society (Wilkinson & Pickett, 2010).

Material Needs

The provision of material basics is an essential prerequisite for learning (Maslow, 1954; Noltemeyer et al., 2021). Pre-existing access to material basics, like a safe, uncrowded home, and access to digital learning materials, influences how effectively students learn during times of crisis. Such access is a function of family economic, social, and cultural capital (Bourdieu, 1984). During a crisis, family capacity to support their children with learning is also conditioned by such capitals and—subsequently—by the capacity of families/carers to navigate the impacts of crisis and disaster themselves. The economic pressure of the COVID-19 lockdowns placed extraordinary financial stress on many families. Three findings discussed in this book are valuable to highlight here.

First, nutritious food is an essential material need for life, and is essential for learning. In Chap. 5 we noted that schools play a vital role in providing food for students from backgrounds of poverty, for example through breakfast clubs and free school lunches. During school closures, students who rely on school for food have increased exposure to hunger and malnutrition, especially if their family is facing additional economic deprivation due to the crisis—such as losing their employment during the COVID-19 pandemic. On return to on-site learning, schools may not have been able to reinstate breakfast and lunch provision due to infection control measures (the wearing of masks, physical distancing, food handling precautions, and avoiding shared use of crockery and cutlery).

Secondly, the COVID-19 pandemic has raised awareness about the deleterious effects of overcrowded and insecure housing on learning, as discussed in Chap. 4. The support schools are able to provide to remedy this is limited in the best of times, focused largely on after-hours homework clubs for students who don't have a suitable study space at home. COVID-19 school closures took away even that small support mechanism.

Third, in Chap. 6 we pointed to material requirements for students to engage with learning from home, including digital hardware and software. Such items are beyond the financial capability of many families—and are typically more easily met by well-equipped school systems. Moreover, internet connections may be unreliable in rural and remote regions, placing students at disadvantage compared to their peers in urban areas.

Across all these aspects of material needs, innovations—borne from necessity during times of crisis—indicate possibilities for action. The response to the COVID-19 pandemic showed how schools and school systems, welfare agencies, and charities and other non-government services overcame barriers that usually make interagency collaboration difficult, to quickly identify students considered vulnerable, with the aim of coordinating responses. During COVID-19, many families received food parcels including recipes. Throughout previous disasters, schools have opened their buildings after hours (and on weekends and holidays) as community hubs and spaces for students to learn. During the COVID-19 pandemic, charities, ICT companies and other businesses provided schools and students (sometimes temporarily) with laptops, dongles, mobile data, and other ICT needs.

Such innovations are worth continuing. In particular, interagency collaboration and partnerships between schools and business are valuable platforms for responding to new hardships as they arise, as well as addressing chronic and ongoing needs of disadvantaged students. As Reimers (2022, p. 470) suggests:

> The efforts to sustain education against the odds [during COVID-19] led to much collaboration within schools, across schools, between schools and other institutions, and across different government entities. These collaborations, intra-education and between education and other sectors, such as health, led to innovations and emphasized that such collaborations are essential to finding ways to address the many and multidimensional needs of students.

However even with such collaborations with other agencies, "education cannot compensate for society"—as Basil Bernstein famously stated in the title of his 1970 article (Bernstein, 1970). The spotlight the pandemic has thrown on the deep economic deprivation faced by many students, even in wealthy countries such as Australia, should prompt more significant poverty alleviation through national and international policies. In the Australian context, the innovative approaches to income support provided in the early stages of COVID-19 lifted scores of families out of poverty—but when these measures ended this "plunged hundreds of thousands of children back into poverty" (Bessell, 2021, para. 18; also Davidson, 2022). In response to such evidence, the CEO of the Australian Council of Social Services (ACOSS) argues:

> The COVID-19 pandemic has taught us that poverty and inequality are not an inevitable state of being. They grow because government policies allow them to, and in many cases, directly increase them. The income supports introduced during the first COVID wave reduced poverty by half and greatly reduced inequality of incomes. [...] Our response to COVID-19 showed we can end poverty. And when we do, it's good for all of us. (ACOSS, 2022, n.p.)

The constituent features of the pandemic's impact and its relationship to educational equity around the world create a unique and specific call to action: children must be lifted out of poverty. The United Nations (2020) has recognised the exacerbation of pre-existing disparities by the COVID-19 crisis, the effects of which have reduced opportunities for many of the most vulnerable children, youth, and adults—those living in poor or rural areas, girls, refugees, persons with disabilities and forcibly displaced persons—to continue their learning (United Nations, 2020, p. 2).

Emotional Wellbeing

Across this book we have emphasised how schools and early learning centres provide children and young people with more than learning opportunities. The impact of wellbeing on learning has long been recognised and is highlighted during times of crisis (Reimers, 2022). The sudden and uncertain nature of crises impacts on emotional wellbeing at the child, family, school, and community levels (see Chaps. 2 and 5). This is both in the immediate event, as well as throughout the recovery and aftermath of the catastrophe. Vital to meeting the emotional wellbeing needs of children and young people is maintaining partnerships between early childhood education and care centres, schools, families, and learners during and throughout the aftermath of crisis and disaster. Two specific aspects are of particular importance in relation to supporting the emotional wellbeing of children and young people during a crisis: access to material basics, and mitigating uncertainty.

During early childhood many issues flow from poverty and precarity, which include lower access to quality nutrition, housing insecurity, and chronic stress. When action is not taken in the early years, impacts accumulate over the life course. Concerningly, long-term follow-up studies of individuals conceived and in utero during previous pandemics, natural disasters, and famines (such as, the Dutch hunger winter of 1944/1945, the 1918/1919 influenza pandemic, the North American ice storm of 1998, and the Chilean earthquake of 2010) show the potential for lifelong negative consequences of such shocks (Yoshikawa et al., 2020).

Experiences during the COVID-19 pandemic foregrounded how this crisis has been interlaced with anxiety, for example due to sudden social isolation, uncertainty about the causes and effects of the illness, and increased economic precarity. Such anxieties have ongoing impacts on both learning and wellbeing. During environmental disasters (like those discussed in Chap. 3) students were faced with immediate and deep fears for their own safety and the safety of their loved ones, and then had to navigate extensive disruption and loss. If students had to relocate to a different school, this created additional emotional pressures, such as building new relationships with peers and staff and adjusting to different routines and expectations. Moreover, the wellbeing of families and staff is also affected by crises, reducing their capacity to support children and young people. These negative impacts on wellbeing created by crises affect many people across the communities, but they have particularly severe consequences for those already struggling and who have the least resources to deal with the additional harm caused by the crisis.

Crises can exacerbate prior concerns about a student's wellbeing. A crisis can disconnect students from the school-based supports they are used to relying on, including their teachers as well as allied professionals such as school nurses, social workers, and psychologists. A particular concern, highlighted in Chaps. 2 and 5, is in relation to students who may not be safe at home due to violence, abuse, or neglect. School staff play an essential (and, in many countries, mandatory) role in identifying children at risk of domestic violence and abuse. However, fulfilling this role is challenging when students' attendance on school sites is significantly reduced due to a crisis. As a result, teachers and allied professionals have less insight into what happens in the home environment and what new traumas students might return with when schools re-open.

Schools and school systems are important sites for supporting student emotional wellbeing throughout and beyond the crisis and disaster. During such events, the pressure on schools to meet the emotional wellbeing needs of students is intensified. School structures and routines, as well as access to trusted adults and peers, are important to emotional wellbeing. Importantly, as discussed in Chap. 3, recovery from trauma and supporting wellbeing takes time. The effects of crisis-related trauma on learning and wellbeing can continue for several years. For younger students, there is the risk that traumatic events can have ongoing developmental impacts including on social and emotional skills. The delayed and ongoing effects of trauma highlight that emotional wellbeing must be a priority for years after a crisis.

Education systems need to support schools and early learning centres to continue their essential wellbeing work long-term; increase access to allied professional staff with appropriate trauma-informed practice skills; and provide workers with professional development to support their students' wellbeing as well as their own. Previous crises have shown the benefits when systems, families, and students to work together to maintain the connection to learning. Interagency collaboration (as noted in relation to material needs) that was significantly enhanced during the pandemic needs to continue, especially in relation to students who are at risk of abuse and rely on both schools and child protection authorities for their safety.

Prior to the pandemic, schools and early learning centres already played a broad role in supporting children's (and families') wellbeing. While recognised in scholarly literature and school system policies, this important contribution was largely invisible to the wider public. The disconnection of students from their formal learning brought about by widespread lockdowns during COVID-19 revealed this vital broader role that schools and early learning centres serve within communities. Working towards enhanced educational equity requires greater visibility and recognition of the role that schools serve beyond learning, as sites of wellbeing promotion (and, as noted above, provider of material goods), especially for already disenfranchised and traumatised children and young people.

Beyond schools and school systems, the COVID-19 pandemic has also underlined the importance of system-wide support for mental health and wellbeing across the community. However, prior to the pandemic, mental health services in many countries were already stressed (e.g., Nicolson et al., 2020) and significant investment is needed to cope with the additional demand as a result of the pandemic. Enhanced use of telehealth and virtual/digital delivery of such services during the pandemic is worthwhile. In addition, when governments implement policies to alleviate economic distress (see section on Material Needs above) this has flow-on benefits for emotional wellbeing, mental health, and physical health (McLachlan et al., 2013).

Finally, the COVID-19 pandemic is a cautionary tale of how, without due consideration in education systems, as well as in broader policy responses, crisis management can result in "shadow pandemics". This term has been used in relation to increased domestic violence (see Chaps. 2 and 4) as well as mental health and wellbeing impacts (see Chap. 3) that have occurred as a result of the COVID-19 pandemic. Khan et al. (2020, p. 235) pick up on multiple meanings of this concept:

> The COVID-19 pandemic and the resultant quarantine and lockdowns have led to a shadow pandemic of mental health disorders, economic instability, racial bias, surge of domestic abuse and spread of misinformation. While the focus has been on the mental health of frontline essential workers in this pandemic, its effects on the mental health of the vulnerable population such as known psychiatric patients, racial minorities, children and elderly have largely been ignored. If we are to prevent a post-pandemic mental health crisis focus needs to be shifted to both include the vulnerable population in the research and to take urgent measures such as provision of online counselling, penalization of spread of misinformation and monitoring hate crime.

For children and young people—especially those already disadvantaged—these negative impacts of the shadow pandemic may prove to be far more serious than the direct health effects of the virus, with long term effects both on their wellbeing and on their learning.

Access to Learning

Pre-existing learning inequalities based on background factors have been well-established in literature over many decades (see Chap. 4). Despite the seeming intractability of educational inequality, there have been promising signs of commitment to change and actual improvement in the twenty-first century (United Nations, n.d.). This book was spurred in part by our concern (and global concern) that the COVID-19 pandemic has led to a significant setback to these advances.

In this section, we focus specifically on the third dimension in our model: access to learning. As is clear from the previous two sections, however; all three dimensions are interconnected and learning, in particular, is a challenge if a child's or young person's emotional and material needs are not met. It is important to acknowledge that—even in the face of challenges created by disasters and pandemics—students are expected to continue with learning. As discussed in Chap. 3, following Hurricane Katrina (2005), the Christchurch Earthquakes (2010–11), and the Great East Japan Earthquake (2011), students were relocated to new schools to continue with their education. Throughout the 1918 influenza and 2009 H1N1 swine flu outbreaks, as well as the COVID-19 pandemic, learning was relocated to the home. As demonstrated through our analysis in Chaps. 3–6, the effectiveness of learning in those contexts is strongly mediated by—and exacerbated by—prior inequalities. Here, we emphasise three key findings.

First, formal learning (in schools and early learning centres) provides the foundation needed for ongoing academic (and lifelong) success, including literacy and numeracy—and students from disadvantaged backgrounds tend to lag on standardised tests in those areas. During COVID-19, learning growth in literacy and numeracy reduced overall, and students from low socio-economic backgrounds were especially negatively affected (see Tomaszewski et al., 2022). Without targeted and substantial support, these learning losses are likely to have deep and long-term consequences. Having fallen behind, on their return to school, many students struggle to keep up with lessons and may, ultimately, disengage and leave school early, casting "a long shadow over the economic well-being of individuals and nations" (Schleicher, 2022, p. 39). Moreover, learning does not only happen at school. Informal learning through play, sport, creative arts, and other leisure activities is also significant. During disaster and crisis, access to many such activities is suddenly cut off. Privileged families have more capacity to make up for this through beneficial informal learning activities at home—enabled by cultural capital, material resources and a spacious home—than disadvantaged families.

Secondly, experiences of learning at home during the COVID-19 pandemic have re-affirmed that student learning occurs best when supported by active teaching by a qualified professional in a face-to-face setting. In the report for its International Summit on the Teaching Profession, the OECD (Schleicher, 2022, p. 3) notes that the pandemic highlighted "that relationships and social interactions in schools are vital for children's and young people's education". As argued in Chap. 6, the provision of content (whether through digital, radio, television, or hard copy materials) is not sufficient to ensure learning—it also requires teachers' professional pedagogical knowledge. The ability of parents and carers to actively engage their children in the learning process was mediated by their own knowledge (including digital literacy), available time, and other pressures. There is a risk that students' love of learning has also been dented, as a result of endless online classes and worksheets that tended to be boring compared to regular lessons. Generally, the quantity (amount of time spent learning) as well as the quality of learning during school closures was much reduced compared to pre-crisis schooling, especially for learners from disadvantaged backgrounds.

Finally, the impact of the pandemic on access to learning is likely to be disproportionately borne by women and girls. For the majority of women and girls, the pandemic increased care demands within households on an unprecedented scale. Many children, especially in developing nations, have faced a higher risk of child labour and lower rates of school enrolment, especially young girls (ILO, 2021). Some of these girls may never return to schooling. When childcare centres closed abruptly along with schools an estimated 1.5 billion children had nowhere to go, nearly one-tenth of the world's population and 800 million of them were under six years old (Zaidi, 2021). Attention to the gendered dimension of impacts of crises on access to learning is important in both developed and developing nations.

Compared to historical crises (see Chap. 3) where policy responses were largely focused on basic needs and crisis containment rather than education, "efforts to generate and fund alternative approaches to educate during the pandemic were significant" (Reimers, 2022, p. 469). There is much to learn from promising innovations during this time as well as from experiences of deepening educational inequity despite policy commitments to schooling.

The COVID-19 pandemic has highlighted potential innovations in the education sector that could improve equitable access to learning (see Chap. 6). Digital learning, provided students have equal access to hardware and software, can ameliorate the once significant boundaries of urban and regional areas and increase student access to learning opportunities. Asynchronous learning can be enabling for some students who have significant other responsibilities, such as work or caring roles. Teachers have substantially increased their own digital literacies and ICT skills, and many schools and school systems have improved the quality of digital learning options. Out of the crisis have emerged opportunities to leverage digital platforms for inclusive learning (Iwabuchi et al., 2022; Zhao & Watterson, 2021). Further, an expanded use of technology to assist with learning has potential benefits for increasing the connections between home and school. This includes opportunities for greater transparency and closer partnerships with parents and carers, provided issues of digital inclusion are addressed at a school systems level.

Calls for additional "high dose" support for students with the largest learning losses (e.g., Azevedo et al., 2022; McDaid et al., 2021; Sonnemann & Goss, 2020) are already reflected in trials of interventions. For example, initiatives in Australia are pairing disadvantaged students with tutors to redress learning gaps during additional, after-school tutoring (The Smith Family, 2021). There are opportunities to learn from these trials to embed more systematic additional individual or small group tutoring support to help reduce educational inequality and ensure students from disadvantaged backgrounds have the foundational skills needed to succeed in school and in life (also see Education Endowment Foundation, 2021).

Improving access to learning in response to the COVID-19 pandemic is getting attention at the highest international levels. To "facilitate inclusive learning opportunities for children and youth during this period of sudden and unprecedented educational disruption" (United Nations, n.d., n.p.), UNESCO established the COVID-19 Global Education Coalition in partnership with other UN agencies, non-government organisations and media and ICT businesses as early as March 2020. Two years later, in March 2022, the Coalition published a brochure which recognises the ongoing and evolving challenges but also has an optimistic message, foregrounding the "unprecedented amount of innovation at every level" that can benefit education globally (UNESCO, 2022, p. 2).

The OECD (2022, n.p.) has pointed out that "Education systems can build on school-led micro-innovations during the pandemic to develop more equitable learning". In its 2021 Education Policy Outlook, the OECD (2021, n.p.) suggests:

> As countries and economies focus on recovering from a profound global crisis and building a better normal, the moment is ripe to imagine a responsive and resilient future for education more fully, and to dare to make it a reality.

To do this, the OECD (2021) proposes a Framework for Responsiveness and Resilience in Education Policy that balances two challenges: the importance of continuing to move towards post-industrial societies and economies, requiring education systems to provide broad skills and knowledge and more individualised learning pathways; and the urgency of developing the resilience in individuals and societies, including in education systems, that is needed to prepare for and adapt to disruptions.

Looking Forward: Putting Learners at the Centre

> We already faced a learning crisis before the pandemic. [...] Now we face a generational catastrophe that could waste untold human potential, undermine decades of progress, and exacerbate entrenched inequalities. (United Nations Secretary-General António Guterres, quoted in UNESCO, 2020, n.p.)

> Never before have we witnessed educational disruption at this scale. Partnership is the only way forward. This [Global Education] Coalition is a call for coordinated and innovative action to unlock solutions that will not only support learners and teachers now, but through the recovery process and in the longer term, with a principle focus on inclusion and equity. (UNESCO Director-General Audrey Azoulay, quoted in UNESCO, 2020, para. 3)

> Even before the COVID-19 pandemic, the world was living a learning crisis. [...] Even worse, the crisis was not equally distributed: the most disadvantaged children and youth had the worst access to schooling, highest dropout

> rates, and the largest learning deficits. [...] The COVID-19 pandemic now threatens to make education outcomes even worse. The pandemic has already had profound impacts on education by closing schools almost everywhere in the planet, in the largest simultaneous shock to all education systems in our lifetimes. The damage will become even more severe as the health emergency translates into a deep global recession. [...] But it is possible to counter those shocks, and to turn crisis into opportunity. [...] As the school system stabilizes, countries can use the focus and innovativeness of the recovery period to "build back better." The key: don't replicate the failures of the pre-COVID systems, but instead build toward improved systems and accelerated learning for all students (World Bank, 2020, p. 5).

These quotes from global organisations exemplify the strong language used worldwide since 2020. The word "unprecedented" has become a catchcry of the pandemic. From the local to the global, there is broad recognition of the gravity of the COVID-19 pandemic for societies in general, as well as specifically for education—and of the need for bold steps to address the "education emergency" (British Expertise International, 2021, p. 4) that has "robbed many students of the opportunities to learn" (Reimers, 2022, p. 474), hitting already vulnerable and disadvantaged learners and their families especially hard.

Crucially, we argue that action is required across all three dimensions of material needs, emotional wellbeing, and access to learning—and at the layers of schools and school systems (as well as societies as a whole) to generate hope for a more equitable future, not only as we emerge from the COVID-19 pandemic, but also so that we are better prepared for inevitable future crises.

Rather than returning to business as usual, societies need to build back better (a widely used phrase, e.g., see Reimers et al., 2022; World Bank, 2020). Core to this endeavour are children and young people themselves. In our model (see Fig. 7.1) the learner is at the centre—however in policy deliberations, scholarly research, and practical interventions the views of children and young people have been largely absent. The Convention on the Rights of the Child (United Nations, 1989) underlines the importance of education to a child's health, wellbeing, and prosperity; as well as the right of children to have a say on matters that affect them. This certainly should apply to educational decision making during and beyond times of crisis. Having started this final section with quotes from people and organisations in positions of substantial power and authority, we end the section, the chapter, and the book, with the voices of children and young people:

> I personally don't think we have really had any kind of voice [during the pandemic]. I don't think we have been depicted as anything. I think we have been depicted more as stats and facts... [we] don't have any sort of representation. (Female, Australia, in UNICEF, 2020, p. 13)

> I think that the president should take actions and thank the children of the U.S. and thank them for the sacrifices that they are making to help stop the spread of COVID-19. We, children, have made many sacrifices for this pandemic, and we have missed our friends, graduations and proms. Sometimes it will feel like we are being punished for this pandemic, when in reality, we are making sacrifices. I think that it would make many children feel appreciated and helpful if our city council members, mayors, governors, and even, the First Family would thank us, children, for our help and sacrifices due to the virus. (Girl, 11, United States of America, in Terre des Hommes, 2020, p. 14)

> I think the government and everybody who's older than us forgets that we're the future... we're the one who are going to make that change later on. We're the ones who are gonna continue on this life and change it... I guess we need to have more conversations and really have our voices heard... nobody's really listening. (Female, Australia, in UNICEF, 2020, p. 13)

> Children are not as clueless as you think they are. They have a voice, and no voice should be shut out. Being a child doesn't mean our voices are less credible, moreover, it should serve as an insight as to what the situation on sectors that you may have forgotten or have not given much importance is like. (Boy, 10, Philippines, in Terre des Hommes, 2020, p. 13)

References

ACOSS. (2022). *Media release 2 March 2022: New ACOSS and UNSW Sydney Report shows how poverty and inequality were dramatically reduced in 2020, but have increased ever since.* https://povertyandinequality.acoss.org.au/news/new-acoss-and-unsw-sydney-report-shows-how-poverty-and-inequality-were-dramatically-reduced-in-2020-but-have-increased-ever-since/

Azevedo, J., Gutierrez, M., de Hoyos, R., & Saavedra, J. (2022). The unequal impacts of COVID-19 on learning. In F. Reimers (Ed.), *Primary and secondary education during Covid-19: Disruptions to educational opportunity during a pandemic* (pp. 421–459). Springer.

Bernstein, B. (1970). Education cannot compensate for society. *New Society, 26*, 344–345.

Bessell, S. (2021). Australia was a model for protecting people from COVID-19 — And then we dumped half a million people back into poverty. *The Conversation*, 12 August 2021. https://theconversation.com/australia-was-a-model-for-protecting-people-from-covid-19-and-then-we-dumped-half-a-million-people-back-into-poverty-165813

Bourdieu, P. (1984). *Distinction: A social critique of the judgement of taste*. Harvard University Press.

British Expertise International. (2021). *Recover learning, rebuild education solutions to the Covid-19 education crisis: A view from UK-based specialist education providers*. https://www.britishexpertise.org/reports/report-education-recover-learning-rebuild-education

Bronfenbrenner, U. (1995). The bioecological model from a life course perspective: Reflections of a participant observer. In P. Moen, K. Lüscher, & G. H. Elder Jr. (Eds.), *Examining lives in context: Perspectives on the ecology of human development* (pp. 599–618). American Psychological Association.

Davidson, P. (2022). *A tale of two pandemics: COVID, inequality and poverty in 2020 and 2021*. ACOSS/UNSW Sydney Poverty and Inequality Partnership, Build Back Fairer Series, Report No. 3. https://povertyandinequality.acoss.org.au/covid-inequality-and-poverty-in-2020-and-2021/

Education Endowment Foundation. (2021). *Small group tuition*. https://educationendowmentfoundation.org.uk/education-evidence/teaching-learning-toolkit/small-group-tuition

Emery, S. (2019). *Cultural wellbeing in classroom communities: A constructivist grounded theory study* (Unpublished doctoral dissertation). University of Tasmania. Retrieved July 15, 2022, from https://eprints.utas.edu.au/31521/1/Emery_whole_thesis.pdf

International Labour Organization. (2021). Building forward fairer: Women's rights to work and at work at the core of the COVID-19 recovery. Retrieved September 29, 2021, from https://www.ilo.org/wcmsp5/groups/public/%2D%2D-dgreports/%2D%2D-gender/documents/publication/wcms_814499.pdf

Iwabuchi, K., Hodama, K., Onishi, Y., Miyazaki, S., Nakae, S., & Suzuli, K. (2022). Covid-19 and education on the front lines in Japan: What caused learning disparities and how did the government and schools take initiative? In F. Reimers (Ed.), *Primary and secondary education during Covid-19: Disruptions to educational opportunity during a pandemic* (pp. 125–151). Springer.

Khan, R. Q., Farooq, Z., & Khan, A. M. (2020). The Corona conundrum: A shadow pandemic of mental health concerns. *Annals of King Edward Medical University, 26*(Special Issue), 235–237. https://doi.org/10.21649/akemu.v26iSpecialIssue.3645

Maslow, A. (1954). *Motivation and personality*. Harper & Row.

McDaid, L., Cleary, A., Robinson, M., Johnstone, M., Curry, M., Ridgway, T., & Western, M. (2021). *What can be done to maximise educational outcomes for children and young people experiencing disadvantage?* (Pillar Report No. 3). Institute for Social Science Research, University of Queensland. https://issr.uq.edu.au/LTC-Pillar3

McLachlan, R., Gilfillan, G., & Gordon, J. (2013). *Deep and persistent disadvantage in Australia.* Australian Government. Productivity Commission Staff Working Paper.

Nicolson, S., Newell, S., & Collyer, B. (2020). *Impacts of COVID-19 on children and young people who contact Kids Helpline (2020).* Australian Human Rights Commission. Retrieved November 16, 2020, from https://humanrights.gov.au/our-work/childrens-rights/publications/impacts-covid-19-children-and-young-people-who-contact-kids

Noltemeyer, A., James, A. G., Bush, K., Bergen, D., Barrios, V., & Patton, J. (2021). The relationship between deficiency needs and growth needs: The continuing investigation of Maslow's theory. *Child & Youth Services, 42*(1), 24–42. https://doi.org/10.1080/0145935X.2020.1818558

OECD. (2021). *Overview. Education Policy Outlook 2021: Shaping responsive and resilient education in a changing world.* https://www.oecd-ilibrary.org/sites/75e40a16-en/1/3/1/index.html?itemId=/content/publication/75e40a16-en&_csp_=afa978e563bc62d2fa53de0de30815a3&itemIGO=oecd&itemContentType=book

OECD. (2022). *Building on COVID-19's innovation momentum for digital, inclusive education. International Summit on the Teaching Profession.* https://www.oecd-ilibrary.org/education/building-on-covid-19-s-innovation-momentum-for-digital-inclusive-education_24202496-en

Putnam, R. (2000). Social capital: Measurement and consequences. *Canadian Journal of Policy Research, 2*(1), 41–51.

Reimers, F. (2022). Conclusions and implications. In F. Reimers (Ed.), *Primary and secondary education during Covid-19: Disruptions to educational opportunity during a pandemic* (pp. 461–475). Springer.

Reimers, F., Amaechi, U., Banerji, A., & Wang, M. (2022). *Education to build back better: What can we learn from education reform for a post-pandemic world.* Springer.

Schleicher, A. (2022). *Building on COVID-19's innovation momentum for digital, inclusive education. International Summit on the Teaching Profession.* OECD. https://doi.org/10.1787/24202496-en

Sonnemann, J., & Goss, P. (2020). *COVID catch-up: Helping disadvantaged students close the equity gap.* Grattan Institute. https://grattan.edu.au/report/covid-catch-up/

Terre des Hommes. (2020). *#CovidUnder19.* https://www.tdh.ch/en/projects/covidunder19

The Smith Family. (2021). *Improving young Australians' literacy and numeracy: The Catch-Up Learning program.* https://www.thesmithfamily.com.au/research/reports/catch-up-learning-pilot

Tomaszewski, W., Zajac, T., Rudling, E., te Riele, K., McDaid, L., & Western, M. (2022). Uneven impacts of COVID-19 on the attendance rates of secondary school students from different socioeconomic backgrounds in Australia: A quasi-experimental analysis of administrative data. *Australian Journal of Social Issues, 00*, 1–20.

UNESCO. (2020). *UNESCO rallies international organizations, civil society and private sector partners in a broad Coalition to ensure #LearningNeverStops*, August 5. https://www.unesco.org/en/articles/unesco-rallies-international-organizations-civil-society-and-private-sector-partners-broad

UNESCO. (2022). *Transforming education through innovation: The Global Education Coalition leading in action*. Retrieved August 19, 2022, from https://unesdoc.unesco.org/ark:/48223/pf0000381023

UNICEF. (2020). *"Living In Limbo". The views and experiences of young people in Australia at the start of the COVID-19 pandemic and national response*. https://www.unicef.org.au/our-work/unicef-in-emergencies/coronavirus-covid-19/young-people-survey-australia

United Nations. (1989). *Convention on the Rights of the Child*. Office of the United Nations High Commissioner for Human Rights. Retrieved April 19, 2021, from https://www.ohchr.org/Documents/ProfessionalInterest/crc.pdf

United Nations. (2020). *Policy brief: Education during COVID-19 and beyond*. Retrieved November 16, 2020, from https://www.un.org/development/desa/dspd/wp-content/uploads/sites/22/2020/08/sg_policy_brief_covid-19_and_education_august_2020.pdf

United Nations. (n.d.). *Sustainable development goals: 4 Quality education*. https://www.un.org/sustainabledevelopment/education/

Wilkinson, R., & Pickett, K. (2010). *The spirit level. Why equality is better for everyone*. Penguin.

World Bank. (2020). *The COVID-19 pandemic: Shocks to education and policy responses*. https://inee.org/sites/default/files/resources/148198.pdf

Yoshikawa, H., Wuermli, A. J., Britto, P. R., Dreyer, B., Leckman, J. F., Lye, S. J., Ponguta, L. A., Richter, L. M., & Stein, A. (2020). Effects of the global coronavirus disease-2019 pandemic on early childhood development: Short-and long-term risks and mitigating program and policy actions. *The Journal of Pediatrics, 223*, 188–193.

Zaidi, A. (2021). *Governments build roads and bridges. Why is another essential piece of infrastructure, childcare, built on women's unpaid labor?* Bill & Melinda Gates Foundation. Retrieved October 14, 2021, from https://www.gatesfoundation.org/ideas/articles/international-womens-day-covid

Zhao, Y., & Watterson, J. (2021). The changes we need: Education post COVID-19. *Journal of Educational Change, 22*(3), 3–12. https://doi.org/10.1007/s10833-021-09417-3

Index[1]

[1] Note: Page numbers followed by 'n' refer to notes.

E. S. Rudling et al., *Education and Equity in Times of Crisis*,
https://doi.org/10.1007/978-3-031-18671-4